Ship Management Series

Shipping Finance

Financing Ships and Mobile Offshore Installations

by J. E. SLOGGETT

B.Sc., C.Eng., F.R.I.N.A., F.I.Mar.E., F.I.C.S., C.Dip.A.F.

We are grateful to Skyfotos for their permission to use their photograph of DYVI TERN on the front cover.

Fairplay Publications

Published and distributed by
FAIRPLAY PUBLICATIONS LTD
52/54 Southwark Street, London SE1 1UJ
Telephone 403 3395
Telex 884595 FPLAY G

ISBN 0 905 045 69 6

Printed in Great Britain by Mayhew McCrimmon Printers Ltd., Great Wakering, Essex and Typeset by JJ Typographics, Rochford, Essex.

The Author

Jolyon Sloggett graduated from the University of Glasgow with first class Honours in Naval Architecture. As a ship design draughtsman with William Denny he was on the team which, in 1959, used the Glasgow University "Deuce" computer to perform ship design calculations, a "world's first". Joining Houlder Brothers & Co as an Assistant Naval Architect he was eventually to become Executive Deputy Chairman of Holder Offshore and Executive Director, Finance and Development for the London based shipping and offshore group, being responsible for the building contracts and financing arrangements for a wide range of vessels. Among his responsibilities during this exciting time when the British shipowner was diversifying into the gas and offshore industries, he had negotiated the contracts and financing arrangements for the *Uncle John,* the world's first semi-submersible MSV. He held a number of appointments on the boards of subsidiary and associated companies, including joint companies with BP and the British Steel Corporation.

In 1978 he was recruited by the newly nationalised British Shipbuilders Corporation as a Managing Director, to take charge of merchant ship marketing and product development, with particular responsibilities for developing the Corporation's activities in the offshore industry.

Since 1981 he has been in his own private practice as a consultant on financial, technical and marketing matters to companies in the UK and overseas. He lives near Steyning, in West Sussex.

To Alexandria and Clementine

Contents

Foreword

by J. G. Davis, Director of Kleinwort, Benson Limited

I am delighted to write some words about this most useful and timely book. After over 10 years of slump the maritime industries, shipowners and shipbuilding are still suffering — perhaps more profoundly than ever and with as yet little prospect of relief — from the malady brought about by the sins of the 50s and 60s.

It has always been my contention that the greatest of the sins of that period was that of ignorance.

None of the participants, shipowners, shipbuilders or, most significantly, banks had a real comprehension of what they were doing during those heady decades which represented the apex of the post-war expansion.

The story has been recounted again and again but the outstanding feature was the "special" financing made available by banks for the building or purchase of ships. The remarkable performance of, above all, the exciting new breed of Greek shipowners — who had very rapidly learned the art of "gearing" in stark contrast to the more conservative shipowners (particularly the UK variety) who continued to pay for their ships in cash from their own resources — was a seductive influence on the very liquid banks who lowered their defences and gave credit — even up to 100 per cent or more — to those borrowers. And since the demand in ships seemed inexhaustible, the banks were for a long period well rewarded and totally untroubled by these high-risk loans, while borrowers in their turn became richer and richer.

But the market changed in 1973 and reality took over from euphoria. I have special reason to remember that year because it was in January '73, eight months before the "crash", that I came into banking from shipowning. It was startling for someone who had always thought of shipowning as a tough way of scratching a living to hear many in the banking fraternity talking of shipowning as a cornucopia of riches. at that time I wrote a piece entitled "A ship is worth what it can *earn*... not even its first historic cost or written-down value". My article was considered by many to be lacking in vision.

These rambling thoughts are perhaps a long-winded way of leading up to this new book. They are, though, very relevant to my appreciation of it.

Jolyon Sloggett's careful analysis will have removed any misconceptions that the reader may have had about shipowning and shipping being a romantic "special case". In other words, ignorance — that cardinal sin — is removed.

The book concerns itseslf more with method than philosophy and with that limitation I warmly commend it as required reading for shipowners and shipping financiers, and the latter might find it useful to make a subsequent study of "Maritime Liens" so that they may thoroughly know just how strong, or how fragile, their security and their mortgage is.

Could I also suggest that Thomas Jefferson's aphorism should be nailed to the shipowner's wall above the bookcase containing this book:

> *"Never buy what you do not want because it is cheap...*
> *it will be dear to year".*

J. G. Davis

Chapter 1

Introduction: The Limited Company

Ships have always represented a major capital investment for their owners. Even the earliest ocean going ships, although small, would have represented a major investment at the time — and a very risky one!

As a consequence of the requirement for large funds the concept of shared ownership of ships, with the profits and losses being shared between a number of individuals, pre-dated the joint stock company by several hundred years. The latter only came into being as a consequence of the Industrial Revolution, when the suddenly very much increased sums required to finance machinery and plant, railways, etc, were beyond the resources of individuals, or even small groups of individuals who might form partnerships based on mutual trust. Finance had then to be obtained from others who would not take an active part in the management of the affairs of the business. Such people could not be expected to accept the unlimited liability for the losses of the business that until then had been accepted by the proprietors of businesses, and so the concept of the limited liability joint stock company was developed.

Until the advent of the steamship in the 1840s, ships were usually owned by the merchants whose goods they carried, but the coming of steam and the change from wood to iron and steel hulls leading to larger and more expensive vessels saw the rise of shipowning as a separate business based on finance raised through joint stock companies — many leading shipping companies of the present day can trace their origins to this time.

Ocean going ships are now virtually all owned by companies and some knowledge of the financial operations of companies is necessary for anyone studying the financial aspects of shipowning.

The joint stock limited liability company is a quite separate legal entity from the individuals who own and control it. The funding and operation of such companies is controlled by legislation in almost all countries, for example in the UK by a series of Companies Acts, in particular those of 1948, 1967 and 1976. More recently there have been the Acts of 1980 and 1981 to implement various EEC directives.

Whereas in a business operated by an individual as sole proprietor, or by a group of individuals in partnership, the liability for making good the losses in the business is unlimited and in the event of a failure might bankrupt the individual,* the liability of the individuals owning a limited liability company which operates a business is limited to the cost of their shares in the company. Thus having purchased their shares in the company the individual shareholders have no further liability provided the shares have been paid for in full.

In order to ensure that the shareholders in a limited company are kept advised of the activities of the company and in particular of its financial position, requirements are imposed upon the Directors of such companies to make reports and present accounts to the shareholders. In the UK such reports and accounts must be presented annually and must be filed with the Registrar of Companies who ensures compliance with the requirements of law. In order to be sure of the accuracy of the accounts they must be audited by independent accountants who add their certificate to them.

When a new company is formed in the UK the Registrar issues a certificate of Incorporation when certain requirements have been met and he has received the company's Memorandum of Association and Articles of Association (unless the Company is not to have the latter). The Memorandum sets out the objects of the company, details of its share capital, etc, while the subsidiary document, the articles, sets out the internal regulations of the company such as the method of appointing directors.

The capital required to enable the company to start in business will be obtained initially by the issue of shares. The Memorandum will give details of particular types of shares issued by the company and the rights attaching to them; over the years many different types of shares have evolved to suit different company and investor requirements.

The 'ordinary' share carries no right to any particular level of dividend, which is determined by the directors in the light of the trading position of the company and the profits remaining after all other liabilities have been met. Similarly if the company goes into liquidation the ordinary shareholders entitlement is to the residue after all other (legitimate) claims have been met. Thus ordinary shares carry the full risks of the business — if successful it is they who benefit from large profits, but if unsuccessful their owners may well lose all the capital that they have put into the business. The ordinary share capital is often referred to as the 'equity' capital and ordinary shares as 'equities'. Ordinary shares usually carry full voting rights in a company but this is not always the case — there are sometimes more than one class of ordinary shares and not all classes may enjoy full voting rights.

Although the ordinary share may have considerable attractions to many investors, there is often a large body of potential shareholders who do not wish to be so fully exposed to the risks of the business but who are prepared on the other hand to forego full participation in large profits. For such investors the 'preference' share has been evolved. This carries a fixed rate of dividend provided the profits are available for the payment to be made. In the event of a liquidation the preference shareholders will be entitled to repayment of the face value of their shares before any payment is made to ordinary shareholders. The dividend entitlement is usually 'cumulative' which means that if a dividend is not paid in any year, then the amount

*See also Chapter 10.

of such dividend is added to that due the next year when the total of the two dividends must be paid before the ordinary shareholders are paid — if profits are not available they are carried on, etc — this will be laid down in detail in the Memorandum. Usually preference shareholders have no voting rights but may acquire them if their dividends fall into arrears.

As an alternative to the issuing of shares, a company can raise finance from the public by the issue of debenture stock, a debenture being a document acknowledging indebtedness on the part of the company. Funds raised in this way are in fact loans and the company is legally obliged to pay the fixed rate of interest that the stock carries irrespective of the level of profits — in the event of it failing to do so, the debenture holders can petition for liquidation of the company. There are three types of debenture — firstly those which are secured against specific assets of the company (and are known as 'mortgage' debentures); secondly those which carry a floating charge over all the company's assets (the most common type); and thirdly those which are unsecured ('naked' debentures or unsecured loan stock). All loans rank ahead of shareholders funds in a liquidation but secured loans rank ahead of unsecured loans. Thus potential investors in the former will look first at the assets secured before making their investment, whereas potential investors in the latter will look first at the company's ability to generate the funds necessary for the interest payments. It goes without saying that secured loans carry a lower rate of interest than unsecured loans because of the higher risk of loss of capital invested in the latter.

Debentures may be repayable on a fixed date, they may be repayable at the company's option, or they may be 'unredeemable' in which case the principal can only be recovered if the company is wound up or other circumstances arise in which the debenture holder can enforce his security. The company may have the option to repay before the fixed date in the case of debentures with a fixed maturity date — an option it may exercise if interest rates fall, replacing the debentures by a new issue at lower rates of interest.

In addition to debenture stock being redeemable as described above, it should be noted that preference shares are also sometimes redeemable in a similar way. In the UK however until the Companies Act of 1981 it was not possible for a company to buy (i.e. redeem) any of its ordinary shares although in other countries e.g. USA this has long been a possibility. The reason for this restriction was originally to prevent the directors of a company which was getting into difficulties from repaying some of the ordinary shareholders ahead of the creditors. Now however a UK company may under certain conditions redeem a part of its ordinary share capital, and it is possible for a company to issue redeemable ordinary shares.

Among the other variations among the different types of shares, convertible shares must be mentioned. Both preference shares and debentures may, and are, sometimes issued giving the holder the right to convert them into ordinary shares under certain conditions at a specified time or times. Such conversion possibilities may make the investment more attractive to the holder.

The choice of type or types of shares to be issued by a company will depend largely on the type of company, the type of investor it seeks to attract, and on taxation considerations, particularly those relating to the payments of dividends and interest on different types of shares.

Having purchased his shares a shareholder has an unfettered right to transfer them unless restrictions are imposed in the company's memorandum and articles. Such a restriction is a characteristic of a private company which was defined in the UK Companies Act as a company restricting the transfer of its shares, having not more than fifty members (shareholders) including employees past and present, and prohibiting any invitation to the public to subscribe for its shares. Any company not meeting this definition was a 'public' company. The 1980 Companies Act in accordance with EEC directives reversed this arrangement and defined the public company, requiring a statement in the memorandum of such companies that they were public companies and requiring the words public limited company (plc) in the company name. Thus it is now possible to distinguish between public and private companies in the UK by their name, in the same way as this is possible in some other countries. The 1980 Act also requires public companies to have a minimum share capital of £50,000. Other companies not complying with the definition become private companies, which no longer necessarily have restrictions on the transfer of shares or the number of shareholders. Nevertheless it is a criminal offence for such a company to offer shares to the public. Private companies are frequently family owned — there were many such companies in shipowning but the number in the traditional maritime nations has fallen steadily for many years principally as a result of taxation.

There is no restriction on companies owning shares in other companies. If a company controls another company — usually by owning more than 50 per cent of the shares — then the owning company is known as the 'parent' company and the owned company as a 'subsidiary' company. If a company owns less than 50 per cent of the shares in a second company (and does not control it), then this second company is known as an 'associated' company. As we shall see in later chapters there are occasions when a company wishing to own a ship will form a new subsidiary company to purchase the ship rather than own the ship directly.

As has already been mentioned, every company must prepare annual accounts for its shareholders, comprising a Balance Sheet and a Profit and Loss Account. We now consider the salient points of these documents which are required by law to provide certain minimum information, although in recent years many public companies have provided very much more than the required minimum. There are currently a number of changes taking place in matters of detail in the accounts of EEC countries as all company accounts within the Community are made to comply with similar regulations. The principles upon which the accounts are prepared are unchanged although the actual presentation will continue to vary considerably from company to company.

The published accounts are intended to provide the shareholders with information as to the sources of the company's capital and the uses to which it has been put (the Balance Sheet) and as to the success of the company in using its resources to generate income through trading (the Profit and Loss Account).

Thus the Balance Sheet is in two parts, one showing the sources of the company's capital in the form of shares, loans, etc., and the other showing how these capital sums have been used to equip the company for its chosen trades and business activities. It shows not only everything that the company owns but also everything that it owes, it shows the position at the close of the last day in the company's financial year (which may be at any time in the calendar year, but will be the same date every year).

Balance Sheets can be arranged in different formats and different terms can be used to describe the individual items. For the purpose of our brief study we shall call one part the 'capital employed' and the other part the 'employment of capital.'

The capital employed section will firstly give details of the authorised share capital (i.e. as detailed in the Memorandum) and then give details of the shares that have been issued, indicating whether or not they are fully paid — shares can be issued partly paid in which case the shareholder is liable to a call for the balance. Any loans will be shown both long term and short term including bank overdrafts.

Suppose then we draw up the Balance Sheet for a new shipowning company which has just purchased its first ship. The capital employed section of the Balance Sheet might look like this:—

Balance Sheet
as at 31st December 19XX

Capital Employed

	£
Share Capital — Authorised and Issued	
1,000,000 Ordinary shares of £1 each	1,000,000
Bank Loan repayable over less than five years at an interest rate of 10 per cent per annum	1,000,000
	2,000,000

The other part of the Balance Sheet could be:

Employment of Capital	£
Vessel, at cost	1,750,000
Stock of consumables (lub. oil, stores etc)	50,000
Cash at bank	200,000
	2,000,000

The reader will readily appreciate that the totals of each side of the Balance Sheet i.e. the Capital Employed and the Employment of Capital must always be equal.

When the company has been trading for a period some additional items will appear in each part. Hopefully the company's funds will have been increased by trading profits which provide additional capital for the company and a source of funds from which future dividends can be paid — for the moment we shall call such sums 'Retained Earnings'.

When the company is firmly established the Capital Employed might look like this:—

Capital Employed

	£
Share Capital — Authorised and Issued	
1,000,000 Ordinary shares of £1 each	1,000,000
Bank Loan repayable over less than five years at an interest rate of 10 per cent	400,000

Retained earnings	700,000
	2,100,000

Similarly additional items will appear under Employment of Capital. In the normal course of trading the company will owe money to suppliers for short periods and the total thus outstanding at Balance Sheet date will be shown under the heading of 'Creditors'. Similar sums owed to the company will be shown as 'Debtors'. The company may have found that it had surplus cash and therefore purchased shares etc in other companies — these will be shown as 'Investments'. The value shown will normally be the price paid for the investments, but in the case of quoted investments the market value on balance sheet date will be shown additionally.

When the company has been trading for any period there is also a further important matter to consider — the assets that have been acquired with the company's capital will probably have a limited life. Thus at some point in the future they will have a very low or nil value, and so arrangements are made to gradually reduce the value shown in the Balance Sheet at a rate which will bring the value down to nil (or the scrap value) at the end of the assets' useful life. Thus a ship might be considered to have a useful life of 20 years and to then have a scrap value of 5% of its initial cost. The amount by which the value is reduced each year is called the 'depreciation', which is charged against the profits in the Profit and Loss Account. It is important to appreciate that this is a book-keeping sum and not an actual cash sum — it is not a sum of money transferred to the bank but an allowance for the fall in value of the asset with time which must be taken into account in considering both the asset value and the profits which are reduced by the depreciation. The reader will appreciate that reducing the value of the assets on the Capital side of the Balance Sheet is thus offset by the equal reduction in the profits or 'retained earnings' on the Capital Employed side.

There are several methods for calculating the depreciation. The simplest, and the one usually used for accounting purposes, is the straight line method by which equal amounts of depreciation are charged each year of the asset's estimated life. For example in the case of the ship mentioned above the annual charge for depreciation would be:—

$$\frac{(100-5)\%}{20} \text{ i.e. } 4.75\% \text{ of the initial cost}$$

The balance sheet would show the initial cost and the total depreciation from date of purchase to Balance Sheet date, the difference being the "book value".

Thus after a few years trading we might have:—

Employment of Capital

Investments		
quoted investments (market value £291,600)	275,000	
unquoted investments	121,000	396,000
Current Assets		
Debtors	134,000	
Stock of consumables	46,000	
Cash at bank	232,375	
	412,375	

Current Liabilities		
Creditors	109,000	
Taxation	30,000	
Proposed dividend	70,000	
	209,000	
Net Current Assets		203,375
		2,100,000

Thus the Balance Sheet shows the position of the company on the last day of its financial year. It does not however show the trading profit or loss that has been generated between the two balance sheet dates and we must turn to the Profit and Loss Account for a statement showing the way in which the company's position has changed during the financial year.

This will begin with the statement of the turnover which in the case of a shipowning company would be the gross revenue earned by the fleet. The next line will probably give the operating profit after deduction of all expenses including depreciation, management costs, etc. Interest payable on any loans will then be deducted to arrive at the profit before tax in the following manner

Profit and Loss Account
for the period 1st Jan 19XX to 31st December 19XX

	£
Turnover	1,100,000
Operating profit	500,000
Interest payable	(50,000)
Profit before tax	450,000
Taxation	(200,000)
Net Profit	250,000
Deduct:	
Dividend on ordinary shares — 7%	70,000
Balance transferred to Retained Earnings	180,000

Notice in the case of the interest payable and taxation the accounting convention of showing a negative number by putting brackets round it rather than adding a minus sign.

During the year the profit being made will result in an increase in the assets of the company. This could be simply an increase in cash at the bank, but more likely the directors will make use of the money by investing it either in the business or in some other way. Whatever the way in which the profits are used, the increase in the book value of the assets will be the same, and provided that the accounts have been properly kept, will equal the balance of the profits transferred to Retained Earnings thus maintaining the equivalence between each side of the Balance Sheet.

The extent to which the company pays out its profits by way of dividend as opposed to investing the profits in the business will be determined by the directors, to whom the shareholders may express their views! In a profitable year some of the profits will almost certainly be retained as a reserve out of which to make up any losses, and perhaps pay a dividend, in any of the following years in which the company's trading does not prove as successful. As a matter of policy the directors may retain further sums to finance the expansion of the business. These retained earnings are frequently labelled 'Reserves' or simply 'Retained Earnings' in the accounts, but different descriptions may be used to differentiate between the two types of retention described above. The former, possibly temporary retention is frequently labelled simply "Balance on Profit and Loss Account" while the more permanent retentions may be described as "General Reserves".

Mention should be made of 'capital' reserves. The reserves/retention described so far are all financed out of revenue and can be classed as 'revenue' reserves. As such they can be distributed to the shareholders (in theory at any rate). However there are certain profits of a capital nature that cannot be distributed by law and are known as capital reserves — an example will be quoted in the next Chapter. Capital reserves are not necessarily shown separately on the Balance Sheet, although this was a requirement in law in the UK until the Companies Act of 1967.

The reader will appreciate a vital point about the reserves shown in a Balance Sheet — they are not sums of money waiting in a bank as a contingency reserve, but rather are profits that have been retained and reinvested in the business. There is sometimes doubt as to whether a particular item of expenditure is revenue, and should appear in the Profit and Loss account, or is capital expenditure and therefore affects the assets shown on the Balance Sheet. This point is considered further in Chapter 8.

Another important point to note is that the total shown for capital employed and under employment of capital is not the value of the company. The former shows the historical sum of the amounts invested in the company from time to time, the latter includes the "book value" of the assets which may be very different from their market value. Some aspects of the assessment of the value of a company are dealt with in Chapter 10.

Even a brief study of company's annual accounts and the accompanying notes can provide a useful guide to the company's general health. One important feature to consider is the availability of working capital, which is the finance required to fund the day-to-day operations of the company as opposed to its fixed assets. In the case of a manufacturing company working capital can be readily visualised as the funds required to purchase raw materials, and to pay the wages and other costs of converting these into finished goods and maintaining stocks prior to sale. Thus a manufacturing company requires substantial working capital to enable it to trade. The position of a shipping company is somewhat different. While some costs will certainly have to be paid in advance e.g. insurance, it may be that hire for the vessels is also paid in advance. Thus, the requirement for working capital in a shipping company will depend upon the nature of the trades in which the ships are employed and in fact some shipping companies are able to operate successfully with a negative working capital. It is therefore important when making inter-company comparisons to compare companies in similar trades.

In accounting terms working capital is defined as the difference between current assets and current liabilities — the Net Current Assets of the balance sheet shown on page 7.

Our final point in this Chapter concerns the value of a share. When a company is formed the directors choose the face value of the shares, known as the par value and this is the price at which the shares are issued. However as soon as the company begins trading the assets represented by each share will change and may come to represent a very different sum — for example a company which regularly retains a substantial part of its profits may multiply the value of the assets per share many times over. If a reasonable return on capital invested is maintained by way of dividend, a 5% dividend may become 50%, which is very misleading and may be a disadvantage particularly when dealing with wage claims! One solution is to have shares with no par value, as is allowed in parts of the USA but not in the UK as yet. The usual solution if the situation presents a problem is to 'capitalise' a part of the reserves, i.e. convert them into shares issued without charge to the existing shareholders — this is purely a book-keeping transaction but does mean that the reserves so capitalised can no longer be repaid to the shareholders (it is very unlikely this would have been the intention in the circumstances). The Balance Sheet before and after the transaction could be:—

Capital Employed

	£
Share Capital — Authorised and Issued	
1,000,000 Ordinary Shares of £1 each	1,000,000
General Reserve	6,500,000
Balance on Profit and Loss Account	1,200,000
	8,700,000

After capitalisation of £4,000,000 of reserves

Capital Employed

Share Capital — Authorised and Issued	
5,000,000 Ordinary Shares of £1 each	5,000,000
General Reserve	2,500,000
Balance of Profit and Loss Account	1,200,000
	8,700,000

The dividend as shown in the Profit and Loss Account would then change from

	£
Dividend on Ordinary shares — 50%	500,000

to

	£
Dividend on Ordinary shares — 10%	500,000

The value of a share as calculated from the assets shown on the Balance Sheet (the 'asset' value) may be very different from the market value which is determined purely by the supply and demand for the shares.

The shares of the larger public companies are usually traded on Stock Exchanges which record and 'quote' the prices at which the shares are traded. In order to get a quotation a company has to comply with the rules of the Stock Exchange which may require that, for example, more information is published about the financial position of the company than is required by law. There are also facilities for buying and selling shares in companies that are not quoted, but in the case of a private company where the sale and purchase of shares is restricted, the market value may be difficult to determine and the sale of shares by a shareholder wishing to raise cash may prove difficult.

Chapter 2

Financing Expansion

A shipping company wishing to expand its fleet can consider a number of alternative sources of finance. Which method is the most appropriate will depend upon the type of company, its existing capital structure, the availability of the different types of finance, and the present and projected profitability of the company. In this chapter we consider the various alternatives.

One obvious method of obtaining funds for expansion is by retaining a part of the earnings as we saw in the previous chapter. This is a sound basis for expansion over a period of time, but in shipping the acquisition of a new ship may well require funds well beyond any retained earnings and other sources must be considered in addition. This is largely due to the steady increase in the size and complexity of modern ships resulting in an increase in cost in real terms.

Additional capital can be raised by similar methods as employed by a new company, such as the issue of shares. However the position is complicated by the existing shareholders whose shares may represent assets well beyond the original cost (i.e. the par value) of the shares. A number of ways have been developed whereby a company can sell additional shares.

In the case of a private company an increase in the share capital may present difficulties unless the existing shareholders are themselves able to provide the additional funds; a private company not being allowed to advertise its shares to the public — also the number of shareholders may be restricted. A private company may therefore be forced to 'go public' by turning itself into a public company if it wishes to expand, although if this course is not acceptable it may be that an alternative can be found — for example in the UK the Industrial and Commercial Finance Corporation Ltd (ICFC) is often prepared to take an equity stake in a company on a temporary basis to facilitate expansion, or the company's bankers may be prepared to advance funds on loans depending upon the existing capital structure.

On paper the conversion of a private company into a public company may appear simple, but nevertheless in practice may take up to a year to accomplish. First the company must be reorganised into a public company with the necessary changes to its Memorandum and Articles. It can then offer its shares to the public through its broker. It is usual to employ a merchant bank or issuing house specialising in new issues to manage the issue and to advise on, and arrange the actual issue.

There are a number of ways in which a new issue can be made.

First there is a private placing in which the shares are offered by the merchant bank to its private clients, frequently institutional investors. This is an inexpensive method appropriate for small issues. A further step from this method is a stock exchange placing where a part of the new issue is placed privately and later the other part is offered to the public on a stock exchange.

Larger issues may be made by advertising to the public at a price which will be almost sure to ensure that all the issue is sold; the company being required to produce a prospectus setting out its past record and future prospects. The issue may be underwritten — that is, if the issue is undersubscribed the underwriters buy the remaining shares. A similar method is the offer for sale, but in this case all the shares are first purchased by the merchant bank and then offered for sale at a small premium. Another method is sale by tender when the public are invited to tender for the shares, the company then accepting the highest tenders downwards until the issue is sold.

In the case of the public company wishing to raise more capital, one course is to issue more shares, which is usually done by offering them to existing shareholders who are given the right to purchase a number of shares in the new issue according to the number of shares they hold e.g. the basis might be 1 for 1. The price set is usually at a discount on the current market price. An existing shareholder can then purchase the new issue or alternatively sell his rights to someone else. It will be seen that such a method of issuing additional shares — a 'rights issue' — is essentially fair to the existing shareholders whether or not they are in a position to purchase additional shares or not. Since the Companies Act of 1980 public companies in the UK are bound by law to first offer shares to their existing shareholders before offering them to the public.

As an alternative to issuing additional ordinary shares, a company may be able to consider issuing debentures and raising loan capital. This will depend upon the existing capital structure and on the state of the market for new issues. The ratios of loan funds to equity funds is known as the 'gearing' and if this is already comparatively high the company will find it difficult to raise more loan funds and may upset the equity shareholders who are carrying the risks. The ease with which loan stock can be sold depends upon the movement of interest rates — long term fixed interest stock is attractive to investors when rates are falling, but not when they are rising.

However since the mid nineteen sixties the position of a shipping company wishing to expand has been made very much easier by the availability of cheap credit provided by governments wishing to sustain their shipbuilding industries. Details of government-backed cheap finance is given in the next Chapter; in theory the borrower should provide adequate security as with a commercial loan, but in the desperate fight for shipbuilding orders in recent years the question of adequate security has often been overlooked in order to obtain shipbuilding contracts and avoid politically unacceptable closures of shipyards. It should also be mentioned that during the boom years of the early nineteen seventies some commercial banks also accepted doubtful security in order to climb onto the shipping loan bandwagon.

There is also another aspect of shipping loans that has a very important bearing on the provision of finance for expansion. The acquisition of a new vessel which is a clearly distinct new asset lends itself readily to a project finance basis in which the finance is provided for the specific vessel rather than the expansion of the company in general. As we shall see, this is the basis of the government subsidised loans and usually also for commercial loans in the industry and is covered in the next chapter.

Another popular method of financing expansion is by leasing rather than purchase of new vessels. By leasing a new vessel the shipowner has nearly all the advantages (and disadvantages) of ownership but without the need to provide the initial capital investment. This method of finance is described in Chapter 9.

Three other methods of raising finance must be mentioned. Firstly there is the possibility of 'sale and lease-back', in which a vessel owned by the company, is sold to a third party (e.g. a finance house) and leased back (i.e. hired) to the former owners for a period of years. The shipping company thus obtains a lump sum immediately, retains the full use of the vessel, but has then to pay hire for the vessel for the period of the lease. Whether or not this is an attractive operation for the shipping company will depend largely upon the way in which such an operation is treated by the taxation authorities in the country in question and upon the shipping company's tax position at the time.

A further method of financing the acquisition of new vessels is by tax savings. Most governments allow accelerated depreciation on ships for tax purposes as will be described in Chapter 8, and as a result the acquisition of a new vessel may save the company substantial amounts of tax that would otherwise have to be paid on profits arising from the company's other vessels or activities. In fact this is only a postponement of a company tax liability and is therefore comparable to an interest free loan from the government. In the case of a UK company prior to 1984 the full cost of either a new or second hand vessel could be set against profits in the year it is acquired and so if the company had large enough profits tax could be saved equal to over half the purchase price of the new acquisition! This of course has been a further encouragement to shipowners to order in good times. It will also be appreciated that in an inflationary situation such as has existed in most countries since the nineteen sixties the ever increasing cost of new ships in monetary terms has enabled many profitable shipowners to avoid tax altogether for as long as they ordered new ships regularly!

As a means of financing new vessels, mention must also be made of government subsidies. These can take the form of subsidies to shipbuilders to enable a lower price to be offered to the potential owner, or can be a subsidy paid direct to the owner to assist in financing the purchase. An example of the first form of subsidy is the Shipbuilding Intervention paid to the UK shipbuilding industry since 1977; the Investment Grants paid to UK owners from 1966-1970 is an example of the latter. Similar subsidies have and do exist in most other shipbuilding and shipowning nations of the world — in some cases the subsidies are openly declared, as in the UK, in others they are concealed from view.

The reader will appreciate from the latter part of this chapter how governments, concerned for their shipbuilding industries have encouraged shipowners to order new ships and expand their fleets by the provision of finance in various ways. That this action has been to the detriment of the shipping industry can hardly be disputed, but the current slump has served as a salutary lesson to owners who fell for the enticements offered by government when earnings were high, burdening themselves with enormous loans without expanding their equity base, with the result that when rates fell they rapidly found that they were unable to pay the capital and interest payments on the loans — less loan funding and more equity funding (which could have been obtained when rates were high) would have alleviated their position. It goes without saying that some such owners are no longer in existence; others have survived through government support.

Chapter 3

Project Finance for New Ships and Mobile Offshore Installations

In considering the sources and arrangements of finance for new ships and mobile offshore units on a project basis we start by considering typical terms of a building contract.

The first important point to note is that the cost of financing the construction is nearly always borne by the shipowners. That is to say, the shipowner is required to pay instalments of the total cost to the builder during the construction period — these instalments usually being arranged to cover the shipbuilders costs as they are incurred. The shipbuilder does not therefore require working capital to finance partly built vessels, this being provided by the owner. A typical pattern of payments could be:

Table 1

Event	Payment (per cent of contract price)
Signing of contract	5
1,000 tons steel delivered to shipyard	5
Commencement of steel fabrication	15
25% steel erected	20
75% steel erected	20
Steel erection completed	15
Launch	5
Delivery	15
	100

As a consequence of this method of payment, in order to protect the position of the shipowner in the event of bankruptcy or similar events preventing completion of the vessel under the contract, all materials and equipment ordered under the contract become the property of the shipowner when they enter the yard. Nevertheless the owner is still very much at risk since a partly built vessel, especially prior to launch, is likely to cost considerably more to complete elsewhere that the sum of the remaining instalments (even if there were no difficulties with labour unions in doing this!) For this reason shipowners have sometimes obtained bank guarantees from the shipyard to cover repayment of instalments in the event of bankruptcy during construction, but full cover of this type is rare. Quite common however is a guarantee covering the first instalment paid on signing since there will clearly be no property passing to the shipowner at this stage.

A ship or mobile offshore unit is essentially a complex structure and provision must therefore be made in the contract for changes in the design and equipment during construction. These changes can be divided into two types:

a) Compulsory changes arising from alterations in the rules of the classification society or statutory authority.
b) Changes required by the Owner as the detailed design is developed or in the light of further experience and thinking.

Provision must be made in the contract for alterations of type (a) to be incorporated in the vessel since otherwise the builder will not be able to comply with the contractural requirements to deliver the vessel with a certain class certificate etc. The manner in which this is done varies from contract to contract.

The calculation of the costs and other effects of changes (e.g. delay to the delivery date) is frequently a bone of contention between owner and builder and there is no doubt that many builders, having accepted a low contract price, have attempted to recover lost ground by adding a large profit margin on their extras. The owner has to rely on the experience of his naval architects and superintendents to detect over-charging; if agreement cannot be reached he can always decide not to proceed in the case of type (b) extras but in the case of type (a) his only protection is the arbitration clause which will be found in virtually all shipbuilding contracts.

It should be mentioned that these charges are referred to as "extras and credits" in the shipbuilding industry because of the extra costs or credits which result to the shipowner. In the offshore industry the expression is not used but reference is made to "change orders".

The total effect on the contract price of these changes is not of course known until delivery of the vessel, and provision must therefore be made for such changes in the financing arrangements. Experienced technical staff can usually make some estimate of the likely total but obviously, particularly with changes of type (a) above, this can be only an inspired guess. In order to ensure that full credit facilities, subsidies, etc are obtained on the increase in price arising from the changes it is common practice to include in the contract price a 'P.C.' sum sufficient to cover the maximum total charge expected.

Another problem which sometimes arises in agreeing a contract price, particularly with sophisticated ships or units for the offshore industry, is that there may not be sufficient information available at the contract stage for the builder to price parts of the work because the design has not progressed far enough. This problem can also be overcome by including a further P.C. sum.

When the vessel or unit is delivered the actual cost of the items covered by each P.C. sum is compared with the sum allowed; if the actual is less then the contract price (which includes the P.C. sums) is reduced by the difference. This adjustment will lose the credit etc on the amount of the adjustment, but the owner does finish up with the maximum allowed, whereas if the adjustment to the contract price is upwards it may not be possible to obtain credit etc on the adjustment. Thus the P.C. sums should always be sufficient to fully cover the final cost.

Almost all shipbuilding and offshore unit construction contracts include provision for the owner to provide certain items of equipment. For example it is very common for shipowners to provide the radio and radar equipment as 'free-issue' to the builder to install as part of the contract. However as credit schemes and subsidies do not usually extend to 'Owners Supply' or 'Owner Furnished' items the tendency in recent years has been to reduce such items to an absolute minimum. Nevertheless it is important to ensure that finance is available to cover everything needed to equip the vessel or unit to enter service and a detailed check of the contract specification should be made to ensure that all outfit items needed to put the vessel into service are clearly specified as either builders supply (i.e. included in the contract) or owners supply. Typical items that may be owners supply are bedding, napery, cutlery, mooring ropes and spare gear, etc, additional to classification society requirements, etc.

In some cases however there may be major items of equipment envisaged as owners supply (or owner furnished), such as containers for a container ship or drilling equipment for a drilling vessel. At the present time such items can usually qualify for credit (and subsidies etc) provided they are included in the building contract price and ordered by the builder (albeit on detailed instructions from the owner and possibly agaist a P.C. sum.

In addition to the items already mentioned the owner will have other costs to pay which most companies would regard as part of the capital cost of the vessel or unit, for example:

(i) Charge by Managers for supervision of construction and expenses.
(ii) Costs of crew standing by vessel prior to delivery.
(iii) Any brokerage payable in respect of the contract which is not included in the contract price.
(iv) In the case of a resale contract (i.e. a contract purchased from another owner) any premium payable to the previous owner and any assignment fee payable to the shipyard.

It goes without saying that if any of these items can be included in the contract price it will be to the owners benefit in that they will then receive credit etc.

There are further costs incurred by the owner in connection with the building contract that have to be provided whether or not they are regarded as capital expenditure:

a) Legal fees re-negotiation of building contract
b) Legal fees re-financial arrangements
c) Other fees and charges regarding the financial arrangements
d) Interest payable during construction on instalments paid to builder.

These items will be considered in more detail later in this chapter. Having considered the costs that will have to be met by the owner arising from his contract we can now examine the sources of suitable funds for financing the project.

Following the tremendous expansion of the world's shipbuilding capacity in the late 1950's and early 1960's the potential supply exceeded the demand for ships. This led governments to take measures to preserve their country's shipbuilding capacity by encouraging owners to build ships. The commonest method employed was to offer long term fixed interest loans at attractive rates. At the same time the average size of ships was increasing rapidly, and thus the average cost of new building; the loans thus provided a very convenient source of finance to shipowners who would otherwise have had to expand their equity base. Inflation from the late 1960's increased the capital required for new construction and further encouraged shipowners to make full use of cheap credit offered.

Fortunately agreement was reached by the governments of nearly all shipbuilding countries on the maximum size, maximum period or repayment, and minimum rate of interest for such loans (the OECD agreement) first made in 1974 and modified in 1980. The agreed figures have been changed on several occasions — at the time of writing this book the maximum amount of the loan is 80 per cent of the contract price; the maximum repayment period is $8\frac{1}{2}$ years and the minimum rate of interest 8 per cent pa. This latter figure is generally interpreted to include the various costs and charges associated with the loan and so the actual rate of interest charged is usually $7\frac{1}{2}$ per cent. This is the case in the UK. The OECD agreement does provide the possibility of variation in the agreed terms if this is necessary to match the terms offered by a competitor from a non-OECD country.

With the great pressure on governments to support their shipbuilding industries since the world shipping slump began in 1975 there has been widespread abuse of the OECD agreement, in some cases quite open and in others concealed. In particular in most shipbuilding countries home owners are offered much more advantageous terms than are offered with export orders. As a result a shipowner or offshore contractor making enquiries for new construction in a number of countries may receive a bewildering variety of different credit terms. The analysis and comparison of such offers is described in Chapter 6, but before any such analysis be made precise details of the credit arrangements are required.

The usual arrangements allow the owner to draw down the loan to assist in the payment of the instalments during construction provided that he makes an initial cash payment of say 10 per cent entirely from his own funds and that thereafter he continues to make payments from his own resources such that the proportion of the instalments paid out of loan funds at any time does not exceed the final agreed loan proportion of the contract price. Thus the instalments required in Table 2 would be paid in the following manner:

Table 2

(All payments no percentage of final price)

Event	Total Payment	Payment from Owners funds	Payment from loan
Signing of contract	5	5	nil
1,000 tons of steel delivered	5	5	nil
Commencement of steel fab.	15	nil	15
25% steel erected	20	nil	20
75% steel erected	20	3	17
Steel erection complete	15	3	12
Launch	5	1	4
Delivery	15	3	12
	100	20	80

The owner must clarify the position as to the payment of interest on the loan drawdowns during construction. This interest may for example be payable every sixth month from the date of the first drawdown with a final payment in respect of the pre-delivery drawdowns on the day of delivery — this has been the standard UK pattern for many years. More recently however there has been a tendency to allow this interest to be 'rolled-up' and added to the total loan outstanding when the vessel is delivered — this substantially reduces the cash sums that the owner has to pay prior to the vessel entering service.

Table 3

Event	Time from signing	Loan drawdown	Interest payment on loans drawdown at 7½% p.a.
Signing of Contract	0 months	—	—
1,000 tons steel del.	6 months	—	—
Commencement of steel fab.	9 months	15	—
25% steel erected	12 months	20	—
—	15 months	—	0.94
75% steel erected	16 months	17	—
Steel erection completed	18 months	12	—
Launch	20 months	4	—
—	21 months	—	2.10
Delivery	24 months	12	1.28
		80	4.32

Thus the cash payments by the owner are increased from 20 per cent to 24.32 per cent by the interest due on the loan drawdowns during construction. But with the possibility of 'rolling-up' this interest and adding it to the loan, the owners cash flow will be much improved, especially as this rolled-up interest will then be financed on very favourable terms! It should however be mentioned that in some cases the rate of interest on the loan is slightly increased if the interest is rolled-up e.g. from 7½ per cent to 8 per cent.

As an alternative to the owner drawing down the favourable loan to make his payments during construction, in some countries the yard itself arranges to borrow the equivalent funds and pay the interest thereon — in this case the contract price is increased to take this into account and the owner is simply required to pay four instalments of 5 per cent during construction, the last being on delivery. This system has operated for many years in French shipyards, and can lead to some involved calculations on delivery if the rate of interest at which the yard has borrowed funds has changed from that originally assumed!

After delivery, repayments of capital in equal amounts and payment of interest are usually made at six monthly intervals, starting six months after delivery. (This is in accordance with the OECD agreement which does however allow the interval to be up to twelve months). The pattern of repayments and interest for an 80 per cent loan repayable over 8½ years with interest at 7½ per cent per annum after delivery (without interest roll-up) then becomes:

Table 4

(all payments as percentage of final contract price)			
Time	**Capital Repayment**	**Interest Payment**	**Total Payment**
6 months after delivery	4.706	3.000	7.706
12 months after delivery	4.706	2.824	7.530
18 months after delivery	4.706	2.647	7.353
24 months after delivery	4.706	2.471	7.177
30 months after delivery	4.706	2.294	7.000
36 months after delivery	4.706	2.118	6.824
42 months after delivery	4.706	1.941	6.647
48 months after delivery	4.706	1.765	6.471
54 months after delivery	4.706	1.588	6.294
60 months after delivery	4.706	1.412	6.118
66 months after delivery	4.706	1.235	5.941
72 months after delivery	4.706	1.059	5.765
78 months after delivery	4.706	0.882	5.588
84 months after delivery	4.706	0.706	5.412
90 months after delivery	4.706	0.529	5.235
96 months after delivery	4.706	0.353	5.059
102 months after delivery	4.706	0.176	4.882
	80.00	27.00	107.00

Following the slump in freight rates from 1975 onwards and the absence of any positive indication of an improvement in the short term, owners contemplating ordering new ships were faced with the prospect that any new ships were unlikely to generate anything like the cash flow required to service the loan repayments in the years immediately after delivery, even if one assumed that eventually the freight markets would improve and justify the ownership of the vessel! As a consequence of this situation some authorities, ignoring the OECD agreement, began offering a moratorium on loan repayments for two or three years after delivery. Such moratoria usually extends to the capital repayments only, the owner being required to pay interest on the loan during the period of the moratoria. The loan repayment periods are thus often extended; in some cases such a credit package was offered by combining the OECD credit with commercial money to provide a moratoria and extend the period of the loan overall but this of course raises the overall rate of interest to well above the OECD minimum (assuming that there is a wide difference between the commercial rate of interest and the OECD minimum).

Also as a result of the fall in freight rates from 1975 some owners got into difficulties over meeting their loan repayment commitments and in the late seventies most authorities arranged moratoria for existing ships where this was necessary.

In the case of loans for mobile offshore units as opposed to ships — see Appendix 3 for the distinction — the amount of loan and the period are usually different from those for ships but otherwise the operation is generally similar. Understandings exist between governments as the maximum 'soft' credit for mobile offshore units — namely 85 per cent of the capital cost, repayable over 5 years.

The foregoing gives a summary of the credit which is usually offered with a shipbuilding contract although many variations are being offered by shipyards desperate for orders and using every possible device to make their offers more attractive to the potential owner.

Just as the actual details of the credit offered vary considerably, so also does the mechanism by which it is offered.

We must first distinguish between 'buyer's credit' and 'suppliers credit'. The former is the case where the loan is made to the buyer of the ship who uses the loan to pay the shipyard which thereby receives the same cash instalments from the owner as if the owner was paying cash for his ship. In the case of 'suppliers credit' the loan is made to the shipyard; the owner then pays the shipyard for the ship over the period of the loan. Nearly all shipbuilding credit is 'buyers credit' — suppliers credit can have tax pitfalls and can also cause problems with subsidies, as will be seen in Chapter 8.

The actual funds for shipbuilding credit almost invariably come from the banks but the way in which different governments provide the interest subsidy varies from country to country, and may also vary according to whether the ship is for a home owner or for export. In some countries the government has set up a state bank or finance house to provide the loans — it may handle ship and offshore unit credit only or it may be broader based and in such a case typically the state organisation will be guaranteed by the government, will borrow from commercial banks at commercial rates of interest, making up the difference between these rates and the low rates at which it lends by using funds given to it by the state. In other countries the commercial banks lend direct to the owner at the subsidised rates, receiving the difference between these rates and commercial rates direct from the government, who also gives its guarantee to the banks.

The UK is an example of a country where the credit is provided by banks to the owner direct, the government's Secretary of State for Industry having powers to issue the guarantees to the banks and to provide the interest make up from government funds, in the case of the UK owners, while the Export Credit Guarantee Department (ECGD) issues the guarantees in the case of foreign owners and the Department of Trade the interest makeup. A brief description of the UK credit schemes is given in Appendix 4 to this book and serves to illustrate the way in which typical credit schemes operate.

A third and popular way of providing ship finance has been through the setting up of ship mortgage banks with government support. In some countries it is not unusual for such banks to take an equity interest in the vessel as well as providing loans of say 50 per cent of the contract price. A further step is the setting up of second-mortgage institutes (with government support) providing further funds against a second mortgage, usually at subsidised interest levels.

In some schemes it is necessary for the application for credit to be made by the owner; in other cases the application has to be made by the builder who should be able to advise the owner as to the procedure to be followed in any case. In spite of the apparent enthusiasm of governments to provide this cheap credit the negotiations and agreement on documentation can be very time consuming and may well take several months. It is therefore a common practice for the owner and builder to sign the building contract but with arrangements that it is automatically cancelled (and any instalments paid are returned) if a financial agreement covering the credit arrangements has not been agreed within a given period e.g. three months. Whatever procedure is followed the owner should be careful to avoid any firm contractural commitment until the financing arrangements are themselves firm, in case he should otherwise find himself with a contract and no financing!

When an owner builds in a foreign country the building contract price and the loan will usually be in a different currency to that in which the owner does his accounting. This raises various problems (and gives rise to opportunities) which are dealt with in Chapter 7.

So far we have not mentioned one vital aspect of loans — namely the security or collateral to be provided to the lender by the borrower in order that the former may be satisfied that he will be able to recover the sum lent. Even in their rush to lend money for new ships governments have not entirely overlooked this aspect of lending.

The prime concern of any bank or finance house in lending money is that the loan will be repaid together with the due interest. When extensive loan funding first became common in shipping the lenders' security requirements were usually satisfied by charges on assets e.g. on the ship being financed. Later however it became more usual to examine the 'track record' of the management of the owning company, and the forecasts of the cash flows from which the loan was to be repaid, although a charge on assets would still be required. There are however other forms of security commonly used, particularly to supplement the security offered by a charge on the ship itself.

Although the acquisition of ships based on substantial long term loans only became the normal practice from the 1960's onwards the borrowing of money against the value of a ship goes back hundreds of years. In sailing ship days the Master of a ship could raise cash to undertake or complete a voyage by pledging the ships hull and gear through a 'bottomry bond' which was repayable when the vessel completed her voyage. If however she was lost the bond was not repayable and consequently the rates of interest charged were very high even though the holder of the bond could insure against the loss of the vessel, although not as easily as in the modern insurance market.

For longer term financing the marine mortgage was developed in the late nineteenth century based on the established land mortgage (with which many readers will be familiar) and on traditional marine practice. Such a mortgage gives the mortgagee (the lender of the funds) the right to have the vessel sold if the mortgagor (the owner) defaults on the loan agreement — by, for example, failing to make due repayments — the mortagee retaining from the proceeds of the sale all outstanding monies due to him, the owner receiving only the residue (if any). Mortgages are normally recorded with the details of the ship in the Register Book at the ship's Port of Registry. It is often possible to effect a second mortgage on a vessel although usually the agreement of the holder of the first mortgage is required under the agreement with him. The second mortgagee would recover his sums due from the residue of the sale money after the first mortgagee had recovered all sums due to him, so that the value of a second mortgage depends very much on the size of the first mortgage relative to the likely sale price of the vessel.

When assessing the security required for a loan on a new vessel, lenders will usually accept a first mortgage on a vessel as being worth 50 per cent or 55 per cent of the contract price provided the vessel is of a type for which there is an established second-hand market e.g. a handy sized bulk carrier. If the vessel is a highly specialised one, e.g. a multi-function support vessel of a new type for the offshore industry, then because of the uncertainty as to its resale value, a first mortgage is likely to be accepted as worth only about 40 per cent of the contract price. A mortgage is an attractive form of security from the owner's point of view since its cost is limited to that of drawing up the mortgage agreement itself. Included in the agreement will be clauses to ensure that the vessel is always properly insured and that the lender has first claim on insurance monies in the event of total loss. Additional security is usually given in the form of bank guarantees, assignment of charter parties, or mortgages on further vessels or other assets. Lenders always prefer bank guarantees, particularly in the case of export finance, where the loan is commonly provided by a bank or banks in the exporting country — the bank guarantees being sought in this case from the owner's own bankers. Where bank guarantees are arranged it is usual for the bank to guarantee the due payment of the initial repayments of the loan, e.g. the first six payments, after which the guaranteeing bank has no further involvement; the lender still having the mortgage as security for the repayment of what is now a substantially reduced loan. A bank guarantee will involve the owner in additional costs. Apart from the legal costs of the necessary agreement the bank will make charges for the guarantee based on the sum guaranteed from time to time. A typical rate for such a charge might be $\frac{3}{4}$ per cent pa, which effectively increases the rate of interest on the first part of the loan by $\frac{3}{4}$ per cent pa.

A more attractive form of security from the owners point of view is an assignment of the vessel's charter party earnings. This is only possible if the vessel is being built to service a profitable long term charter from a major company of undoubted financial strength or backing e.g. a state corporation or a major oil company. With such an arrangement the charter hire is paid by the charterer to the lender who deducts the due loan repayments (or a monthly proportion thereof) and pays the owner the balance. In this way the lender is assured that the vessel's earnings are used first to repay the loan. The main value of a good long term charter is to demonstrate the ability of the owner to repay the loan — it gives the lender confidence in the owners cash flow projections. Some major fleets have been built up on the basis of vessels built to service long term charters which have used as collateral to obtain the necessary finance.

In some circumstances an owner may find it convenient to offer a mortgage on an existing vessel as additional security. This mortgage could be either a first mortgage on a vessel otherwise mortgage free, or, perhaps more likely, a second mortgage on a vessel which already has a first mortgage. Such a second mortgage can in some cases be quite valuable — for example a six year old vessel which would have 25 per cent of its original loan outstanding representing 20 per cent of the original contract price, but with inflation over six years this might be only 10 per cent of the current rebuilding price so that the difference between the first mortgage covering the outstanding loan on the vessel and the market value of the vessel, as represented by the second mortgage could be a useful sum, depending of course upon the state of the market for such ships.

In some countries it is also the usual practice for the banks to require personal guarantees from senior executives of the owning company, e.g. the chairman and the chief executive, the banks thus ensuring that these officials take a very real interest in the vessel's success — the actual value of the guarantees probably being relatively small in the event of a bankruptcy, and unlikely to be pursued by the banks unless they feel the individuals had been reckless or negligent.

Lenders will usually be prepared to consider any other security which can be seen to safeguard their position. For example in the case where the owning company is a subsidiary of a large group of undoubted financial strength the lender may well accept a guarantee from the parent company in lieu of a bank guarantee.

It should be mentioned that in their fight for orders there have been cases in recent years when shipbuilders, rather than lose a potential order because the owner could not satisfy the lenders' security requirements, have themselves provided the necessary guarantee! There have also been cases where the shipbuilders have become involved in joint venture leasing companies etc. all designed to facilitate the securing of orders but quite without normal commercial practice.

During the construction phase the lenders will require security for the loan drawdowns. We have already seen that the partly built vessel together with all equipment and materials for her at the property of the owner during construction (p.15) and so the owner can give the lender a legal charge over the vessel etc. However this has limited value in the early part of the building programme and the lender may well insist on the builder providing financial guarantees as already discussed (p.15).

One important point should be emphasised. Although we have talked of financing based around the particular project i.e. the construction of a new vessel, the liability of the owner to the lender is not restricted in any way to that particular project but embraces his whole business. Thus the provision of security as outlined above is to protect the lender in the event of the owner's business as a whole failing; the owner entering into a legal obligation to repay the loan with interest irrespective of the success or otherwise of the particular project for which the finance is provided.

One way in which an owner could limit his liabilities to the particular project would be to put the project in subsidiary company with no other business, so that in the event of failure that subsidiary could be allowed to go bankrupt while the owners other business continued. In such circumstances the lender will see this possibility and probably require a parent company guarantee thus nullifying the effect of the move so far as he is concerned. (There are however other reasons for setting up 'one-ship' companies as will be seen in Chapter 5).

Most lenders require cross-default clauses in their agreements whereby a default by the owner on any other loan can be regarded as a default on the loan in question. This ensures that the lender cannot be put at a disadvantage relative to other creditors when the owner gets into difficulties because he cannot act until there is a default on his particular loan. The consequences of cross-default clauses can be very serious in a large group where there may be a number of subsidiaries with their own loans — default on one loan in a subsidiary could, through cross-default clauses, result in all loans throughout the group being in default, with possibly disastrous consequences. The possible effect of cross-default clauses must always be carefully considered.

Having now described the broad principles upon which government backed credit schemes for new ships operate, there are some important matters of detail which must be mentioned.

Firstly, all legal fees in connection with the agreement, mortgages, etc are almost invariably for the owners account and can now be substantial especially where export credit is concerned. As the use of new construction loans has extended dramatically in recent years so more and more contingencies have been visualised by lawyers as requiring appropriate clauses in agreements which have become longer and more expensive as a result. Legal costs have therefore risen steeply over recent years.

There are also other fees and charges to be paid by the owner during the setting up of the arrangements and prior to delivery. A payment is usually required to secure the government guarantee — this may be for either owners or builders account, usually the owner being required to pay for home owners scheme guarantees and the builders for export schemes. The amount to be paid can vary substantially especially for export guarantees and is typically from 1 per cent of the loan guaranteed to 4 per cent or even more. This is one reason why the price of a ship offered with credit terms may be somewhat higher than the price to a cash buyer (or to one who arranges his own (commercial) credit). Because these charges for guarantees can be substantial it is sometimes possible to pay them by instalments during the building period.

Further fees and charges that are usually borne by the owner are those charged by the lender under two headings — commitment fee and negotiation fee. The latter needs no explanation and is relatively small, if indeed one is charged. The commitment fee is a charge by the lender for holding funds available until needed by the borrower from the time that the lender is committed to making the loan until the loan has been fully drawndown by the borrower. It may be charged at a certain rate per annum on the undrawndown amount, calculated in the same way as an interest charge, or it may be charged as a flat fee when the loan agreement is signed.

One point that should be mentioned in connection with government supported finance is that because the basic reason for the provision of the loans is often not only to help that nation's shipbuilding industry but also the supporting industries (e.g. marine engineering) there may be restrictions on the use of imported materials and equipment particularly where home produced alternatives exist. This restriction may be important where extensive specialised equipment is fitted on the vessel e.g. on a drill ship. The consequences of exceeding the imposed limitations are usually a reduction in the size of the loan. A possible solution may be to finance the hull with the usual shipbuilding credit and to finance the equipment with export credit from the supplying country thus maximising the amount of credit that can be obtained.

A further matter for this chapter is a brief description of promissory notes which are sometimes required to be issued by the owner under the terms of the loan agreement. A promissory note can be defined as an unconditional promise in writing made by one person to another signed by the maker, engaging to pay on demand or at a fixed or determinable future time, a sum certain in money to, or to the order of, a specified person or to bearer. In some loan agreements the owner is required to issue such a note or notes in respect of each drawdown thus acknowledging his liability to make certain repayments on certain dates. Because promissory notes are negotiable the lender can then sell them to obtain funds — the method is commonly used where the credit is given by a state bank or financial organisation, but is not usually required where commercial banks lend direct to the owner.

Finally mention should be made of what might be regarded as the ultimate ship credit — the use of government aid. Many of the worlds industrial nations give 'aid' funds to developing countries and on occasions ships may be included in the package. As can be imagined the possibility has led to ingenious schemes by owners involving the setting up of companies in developing countries to order ships to be ultimately used by the real owner resident elsewhere. There are however situations where, because such aid is usually on a government to government basis, the shipowner in the developing country may be required to pay his government on similar terms as those he might get under a normal shipbuilding agreement; he therefore gets no benefit from it and will not see it as an advantage.

To conclude this chapter it is worth listing the various aspects of loans that need to be established to enable proper financial appraisal to be made. The preliminary essentials are:

1. Amount
2. Period of loan from delivery
3. Rate of interest
4. Currency of loan
5. Repayment details. i.e. is capital repayment in equal instalments at six monthly intervals starting six months after delivery?
6. Is interest calculated and paid in six monthly intervals in arrears?
7. Can loan be drawn down during construction?

Further information required to make a detailed appraisal would include:

8. What initial cash payment is required from owner?
9. What restrictions if any apply to rate of loan drawdown during construction?
10. Details of all fees and charges payable by owner in connection with the loan.
11. Can interest on loan drawdowns during construction be rolled up and added to capital sum at delivery? Does this affect rate of interest payable?
12. What security is required?

Chapter 4

Financing New Ships and Mobile Offshore Installations — Commercial Funds

Since the collapse of the market for new ships in the mid nineteen seventies, the governments of virtually all shipbuilding nations have ensured that cheap credit has been available for orders obtained by the shipbuilding industry. Commercial loans i.e. loans provided independently of any government assistance have therefore been relatively few and have been mainly as additions to government subsidised loans on new ships or for second hand ships.

Prior to the mid nineteen seventies not all shipbuilding nations provided cheap credit — some had no need to because they could obtain all the work they required without such assistance — and commercial loans were therefore very much more commonplace.

The big advantage that a commercial loan has over a government scheme is its flexibility. The latter schemes all follow a fixed pattern of drawdowns and repayments, whereas in theory there are no restraints on a commercial loan. The banks are after all, in business to lend money and provided they can be satisfied that they will be repaid according to an agreed schedule, will be very flexible in agreeing to the arrangements.

The principles upon which a commercial loan operates are generally similar to those described in the previous chapter, the owner being required to satisfy the lender of his ability to repay the loan with reference to his track record and to cash flow projections for the future. Adequate security will be required in a similar way as already described. It is worth noting that in the boom in the early seventies many banks anxious to climb on the shipping bandwagon, were prepared to lend against a first mortgage only, paying little attention to track record or cash flows — when the boom ended and ship prices plummetted such lenders came to regret that they had overlooked basic principles of banking.

Another major difference between the subsidised loan and the commercial loan will be in the rates of interest. The former are fixed whereas the latter are usually variable, and at a higher level.

The interest rates are usually fixed either in terms of the lending banks published base lending rate or in terms of an established market rate, usually LIBOR (London Inter-Bank Offer Rate), with, in each case, an addition that is fixed and incorporated in the loan agreement. Where the bank's base rate is used the rate payable by the owner will change as the base rate changes; where LIBOR is used the rate is usually changed at fixed intervals e.g. six monthly. These arrangements are however entirely for negotiation between lender and borrower.

Another way in which commercial loans differ from many subsidised loans is that rather than the lending being from one bank (as it often is where a government guarantee is given) it is usually from a syndicate of banks. One bank will take the lead and undertake the negotiations on behalf of the syndicate which in the case of large loans might include upwards of twenty banks from a number of countries. From the borrower's point of view the fewer the number of lenders involved the better since this will tend towards quicker negotiations. This may be of vital importance if it should become necessary or desirable to renegotiate the arrangements at a later date. Furthermore a large number may well include one or two banks not so familiar with the type of business involved and who may wish to terminate the loan should difficulties arise even though the more experienced lenders may be happy to continue. The successful conclusion of the arrangements for a large syndicate loan is often marked by the publication in the financial press of 'tombstones' (advertisements) detailing the borrower, the details of the loan, and listing the lenders.

Because there are no governmental or other agreements restricting the terms of a commercial loan (apart from ordinary legal requirements), it may be possible to incorporate a moratorium on capital repayments, or reduced capital repayments, in the first years after delivery. Alternatively one could consider a loan repayable in such a way that the total of capital and interest was equal for each repayment throughout the life of the loan rather than having equal capital repayments with declining interest payments as a consequence as shown in Table 4 (page 20).

Some shipyards have used commercial loans in combination with subsidised loans to try and produce a more attractive proposition for a potential owner. For example a commercial loan can be used to extend the period of a subsidised loan at the cost of an increase in rate. Suppose we start with an 80 per cent loan repayable over 8½ years with interest at 7½ per cent pa, and use commercial funds on which interest is payable at a fixed rate of 15 per cent pa to extend the period of repayment to 12 years. The combination would produce an 80 per cent loan repayable over 12 years with interest at 8.85 per cent pa — if the commercial rate of interest was 10 per cent pa rather than 15 per cent pa then the rate of interest on the combined loan would be 8.02 per cent rather than 8.85 per cent.

Alternatively the owner might ask that the repayments in the earlier years of the loan should be reduced in the belief that inflation and improved rates would enable him to make much larger repayments in the later years. This requirement could be met as mentioned above by equal repayments representing capital plus interest throughout the period of the loan. (Many readers will know that this is the usual method used for house (mortgage) loan repayments!) For the combined loan with commercial money at 15 per cent p.a., the six monthly payment would be 5.663 per cent of the contract price for twelve years; the equivalent rate of interest would be 9½ per cent p.a. This should be compared with the six monthly payments shown in Table 4 for the subsidised loan.

There are various ways in which such a combination of loans can be operated. In order to visualise what is happening let us assume that the commercial lender makes all the necessary payments on the subsidised loan; the difference between these payments and his receipts from the owner being treated as drawdowns or repayments (as the case may be) on his notional loan to the owner, whose payments in respect of both loans will be made to the commercial lender in the form of repayments on a single 'combined' loan. One method for operating such a transaction is for the lending bank to form a subsidiary whose sole purpose is to receive the loan funds as required and make the necessary repayments.

A method for calculating the effects of combining cheap loans and commercial money is given in Appendix 2; the following table compares the subsidised loan with one of the alternatives outlined above. For convenience it has been assumed that the loans are drawndown on delivery; all figures are percentages of the contract price.

Table 5

Year (from delivery)	Subsidised loan repayments	Combined loan: Capital repayments	Combined loan: Interest payments	Combined loan: Total payments	Commercial loan: Drawdown repayment	Commercial loan: Interest on capital	Commercial loan: Loan outstanding	Total loans outstanding
	(1)		(2)		(3)			(4)
1	7.706	3.333	3.538	6.871	0.835	—	0.835	76.129
	7.530	3.333	3.391	6.724	0.806	0.063	1.704	72.292
2	7.353	3.334	3.244	6.577	0.775	0.128	2.607	68.489
	7.177	3.333	3.096	6.429	0.748	0.196	3.551	64.727
3	7.000	3.333	2.949	6.282	0.718	0.266	4.535	61.005
	6.824	3.334	2.801	6.135	0.689	0.340	5.564	57.328
4	6.647	3.333	2.654	5.987	0.660	0.417	6.641	53.699
	6.471	3.333	2.506	5.839	0.632	0.498	7.771	50.123
5	6.294	3.334	2.359	5.693	0.601	0.583	8.955	46.601
	6.118	3.333	2.212	5.545	0.573	0.672	10.200	43.140
6	5.941	3.333	2.064	5.397	0.544	0.765	11.509	39.743
	5.765	3.334	1.917	5.251	0.514	0.863	12.886	36.414
7	5.588	3.333	1.769	5.102	0.486	0.967	14.339	33.161
	5.412	3.333	1.622	4.955	0.457	1.075	15.871	29.987
8	5.235	3.334	1.474	4.808	0.427	1.190	17.488	26.898
	5.059	3.333	1.327	4.660	0.399	1.312	19.199	23.903
9	4.882	3.333	1.180	4.513	0.369	1.440	21.008	21.008
	—	3.334	1.032	4.366	(4.366)	1.576	18.218	18.218
10	—	3.333	0.885	4.218	(4.218)	1.366	15.366	15.366
	—	3.333	0.738	4.071	(4.071)	1.152	12.447	12.447
11	—	3.334	0.590	3.924	(3.924)	0.933	9.456	9.456
	—	3.333	0.443	3.776	(3.776)	0.709	6.389	6.389
12	—	3.333	0.295	3.628	(3.628)	0.479	3.240	3.240
	—	3.334	0.148	3.482	(3.482)	0.242	0.000	0.000

Notes: (1) See Table 4 for details
(2) Calculated at 8·846% (see Appendix 2)
(3) Difference between columns 2 and 5
(4) Total of capital outstanding on subsidised loan plus column 8

The last column in the table shows the total capital outstanding at six monthly intervals, which the commercial lender will wish to consider carefully in relation to the security which is held. It is worth noting that when the subsidised loan has been paid off after 8½ years, the amount of the commercial loan outstanding is 21 per cent of the contract price; the alternative repayment pattern of equal total instalments proposed earlier was adopted this capital outstanding would rise to approximately 30 per cent of the contract price, showing the need for additional security for this pattern of loan repayments.

Another feature sometimes incorporated in commercial loans is the 'balloon' payment at the end of the loan. Thus instead of arranging a twelve year loan with equal capital repayments of 4.167 per cent of the loan each six months, the loan might be for the same sum with the same capital repayments each six months, but at the end of say the eighth year, a repayment of all capital still outstanding would be due, and the loan terminated. Such a payment is known as a balloon payment. When the time came for the balloon payment the owner would arrange to refinance it according to the circumstances at the time. The advantage of such an arrangement is that the owner gets the benefit of the same level of repayments as if on a long term loan, while the lender is not exposed to the level of interest rates far into the future if he is quoting fixed rates and may not be able to cover himself for the full term. He is also protected against changing circumstances.

The sources of loan funds for shipping usually come from the banking system. There are however other possible sources which should not be overlooked and which have been used successfully on occasions in the past when subsidised loans were not available. Life assurance companies have a requirement to invest funds long term and accept fixed rates of interest, likewise the large pension funds have considerable funds to invest. Such funds have been obtained in the past by the issue of debentures; in another well known case the owning one-ship company was structured round preference shares held by such investors — the ship having a long term charter from a state corporation.

These institutional sources of funds have been somewhat overlooked since subsidised loans became available in virtually every shipbuilding nation, but there are still occasions where this use may be advantageous especially for example, by established companies in the offshore industry. It should also be remembered that many institutional investors are also willing to consider equity investment where the project looks attractive.

Another possibility for longer term fixed interest funding which has recently become available is by use of the financial futures market (e.g. LIFFE, the London International Financial Futures Exchange). It will be interesting to see the extent to which lenders make use of this to find sources of funds for shipping loans.

It should also be noted that in some countries e.g. Germany, it is common for commercial banks lending on ships to take an equity stake in the vessel in addition to the loan. In other countries e.g. the UK this is regarded as very exceptional practice.

The taking of an equity stake in a ship is facilitated by the possibility in most countries, of registering a ship as having more than one owner. Each owner then bears his share of all profits, loans, liabilities, etc., of the vessel in the manner that was common prior to the introduction of the limited company. In the UK a ship is divided into sixty-fourths under the terms of the Merchant Shipping Act of 1894—each sixty fourth can have up to five owners registered. The use of this possibility as a possible means of facilitating ship finance is discussed in Chapter 10(2) of this book.

Chapter 5

Financing Second Hand Vessels and Conversions

When considering the financing arrangements for a second hand vessel there are several possibilities to consider.

Firstly the vessel may already have a loan attached which would normally be repayable at the time of sale. Sometimes however the outstanding loan can be transferred with the change of ownership provided the necessary security can be provided and usually provided there is no change of flag — this arrangement usually applying to transfers where the loan has been arranged under a 'home owners' scheme as opposed to an export credit scheme. If an owner has a new vessel constructed with the possibility of an early sale in mind he would be well advised to consider placing the ship in a one-ship company — when the time of sale comes the company rather than the vessel can be sold and this usually facilitates the preservation of the subsidised loan, which may be attractive to buyers and thereby add a premium to the market value of the vessel. There is also a possible disadvantage to the proposal which is discussed in Chapter 10(4).

If the sale of the vessel is to another flag it is possible that new subsidised credit may be available under an export credit scheme. For example, in the UK if a British built vessel, having been owned by a British owner, is then sold overseas it is regarded as an export, and export credit can be arranged. The principal disadvantage with this scheme is the time taken to make the necessary arrangements after the buyer has been identified (see Appendix 4) — ship sales usually take place over a relatively short timescale and this is probably why little use has been made of this facility.

If neither of the above possibilities exist, the new owner must rely on commercial finance. The extent to which lenders will be prepared to support the purchase of a second hand vessel will depend in the first instance on their being satisfied that the owner will be able to meet such obligations to them as may be arranged. As with a new vessel, the owners track record and cash flow projections will be examined. Security will be required on similar lines as previously discussed; the value placed on a first mortgage — vis-a-vis the purchase price will be very much a judgement by the lender based on the level of market value at the time as determined by the freight market. Other details of the loan will be generally as already discussed for commercial loans on new ships.

Conversions are usually financed with commercial money where the costs of conversion lead the owner to seek additional finance. Some short term finance can sometimes be arranged with the shipyard/repair yard in the form of delayed invoicing but for a matter of months only. The ease with which a commercial loan can be arranged will depend upon the usual factors already discussed e.g. ability to repay the loan, security etc.

In some countries it is possible to obtain government subsidised loans for major conversions — for instance such a scheme was introduced in the UK in 1979 for UK owners undertaking the conversion in UK shipyards. This scheme is run on similar lines to that covering subsidised loans for new ships and applies to ships of 1000 gross registered tons and over, and to mobile offshore installations weighing not less than 5,000 tons excluding fuel and water. The rates of interest are the same as for new ships and for new mobile offshore installations, the amount covered by the loan can be up to 80 per cent of the conversion contract value for ships or 85 per cent for mobile offshore installations; the period of the loan is a maximum of five years in each case.

Of course one way of buying a second hand ship is to purchase the owning company! There is however a disadvantage in purchasing ships in this manner, which will be considered further in Chapter 10(4). If a company is purchased as a going concern in order to acquire its ships at an attractive price there may also be consideration relating to the existing staff etc., which may reduce the benefit.

Chapter 6

Financial Appraisal of New Projects

Having considered the manner in which the acquisition of a vessel or unit can be financed we can turn to the methods used to compare alternative projects with one another or to measure the value of a project against a particular yardstick in financial terms.

Three important points must be emphasised before considering the methods commonly used. Firstly the value of any analysis can only be as good as the information used. Only in comparatively rare cases can the cash flows generated by the proposed project be forecast with reasonable certainty. In most cases the forecasts are only estimates and so any analysis can only be as good as such estimates. This must never be forgotten.

The second important point that must be made is that financial analysis and indeed financial considerations as a whole are only one of the factors that a board of directors will take into account in making investment decisions. Thus although the methods of financial appraisal described in this chapter form invaluable tools in assisting management to make the most advantageous investment decisions, it must not be thought that the results of these methods should be in any way the sole determinant.

Thirdly it should be borne in mind when presenting the results of financial analysis to a board of directors that probably only a minority of them have had any formal financial training. It is therefore important that the results can be readily understood by those without detailed financial knowledge, which requires a simple and clear presentation of the results of the analysis.

There are a wide range of methods of appraisal in use; we shall describe three only — two providing a quick and sometimes useful guide while the third has come to be accepted as probably the principal method of appraisal over a wide range of industries.

All the methods have one thing in common — they are based on the 'net cash flows' generated by the project. Such cash flows represent the change in the company's cash receipts and expenditures over a given period arising as a consequence of the implementation of the project. It is important to appreciate that we are talking of *cash* receipts and expenditures; accounting charges and provisions such as depreciation are not included since we are considering strictly the flow of money only. Such flows include all movements of money whether for capital expenditure (and receipts) or trading expenditure and income. To facilitate the calculations, it is assumed that the net cash flow for each period occurs at the end of that period.

The first, very simple method, is the 'pay-back' which is still widely used in many businesses, and forms more a measure of the liquidity (i.e. the flow of cash) than of profitability. The 'pay-back', as its name implies, is simply the number of years required for the cash generated by the project to equal the original outlay. We can consider a simple shipping investment as an example. Suppose a second hand vessel is purchased for $3 million. It is anticipated that after all operating and management expenses etc. have been paid its trading will generate a net cash flow of $1 million pa for five years, after which it will be sold as scrap for $500,000. The net cash flow resulting from the project will therefore be as set out in Table 6; note that the start of the project is denoted as year 0; the first year's trading as year 1, etc. When the cumulative cash flow equals zero the pay back point has been reached.

Table 6

Year	Net cash flow — Case I	Cumulative cash flow
0	(3,000,000)	(3,000,000)
1	1,000,000	(2,000,000)
2	1,000,000	(1,000,000)
3	1,000,000	0
4	1,000,000	1,000,000
5	1,500,000	2,500,000

Thus the pay-back period is three years. However the net cash flows could follow a very different pattern and still have the same payback period — two such examples are shown below which illustrate the two disadvantages of the method — its failure to consider the pattern of the cash flows within the payback period, and its failure to consider the cash flows following the payback period.

Table 7

Year	Net cash flow — Case II	Cumulative net cash flow	Net cash flow Case III	Cumulative net cash flow
0	(3,000,000)	(3,000,000)	(3,000,000)	(3,000,000)
1	250,000	(2,750,000)	1,000,000	(2,000,000)
2	750,000	(2,000,000)	1,000,000	(1,000,000)
3	2,000,000	0	1,000,000	0
4	1,000,000	1,000,000	2,000,000	2,000,000
5	1,500,000	2,500,000	2,000,000	4,500,000

Thus the payback method has obvious disadvantages as a method of analysis. Nevertheless there are two situations where it may be found of use. The first is the case where there is the possibility of sudden collapse of trading income, e.g. a ship built with a good charter for some years from delivery, but with completely unpredictable possibilities thereafter.

Use of the payback method can show whether or not the capital cost will have been recovered when the point at which trading might cease has been reached.

A second use for the payback that the author has found useful in the shipping industry is in fact a graphical representation of the cumulative net cash flows for a new project — this cumulative figure will equal zero at the end of the payback period. This method shows the build up of the investment in the vessel during the construction period and its subsequent reduction after delivery. The graphical representation overcomes some of the objections to the payback method since the year-by-year position can be immediately seen. It has also been found helpful to plot the cumulative net cash flow allowing for an interest charge on the sum outstanding each year, the rate of interest being either that appropriate to an alternative safe (risk free) investment (e.g. gilt-edged stock) or at the company's cost of capital (which will be discussed later in this chapter). A typical graphical representation of a shipping project is shown in Diagram 1. It will be readily appreciated how well this presentation meets the third point made at the beginning of this Chapter.

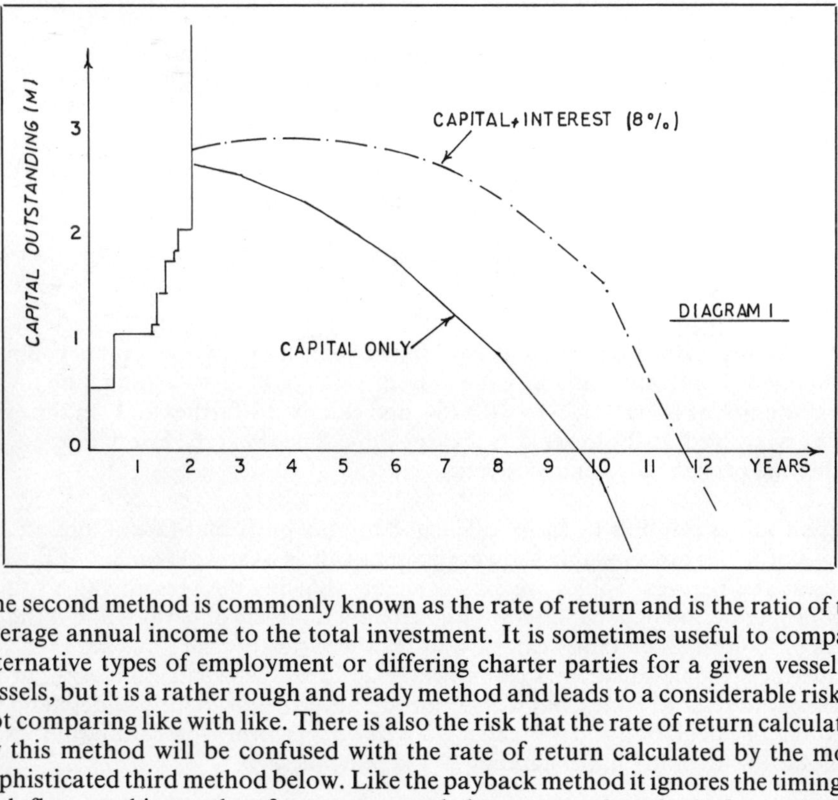

The second method is commonly known as the rate of return and is the ratio of the average annual income to the total investment. It is sometimes useful to compare alternative types of employment or differing charter parties for a given vessel or vessels, but it is a rather rough and ready method and leads to a considerable risk of not comparing like with like. There is also the risk that the rate of return calculated by this method will be confused with the rate of return calculated by the more sophisticated third method below. Like the payback method it ignores the timing of cash flows and is not therefore recommended as a general method of appraisal.

To illustrate the method we can use the project already described. The rate of return in each of the cases would be:—

$$\text{Case I} \quad \frac{5{,}500{,}000}{5} \times \frac{100}{3{,}000{,}000} = 36.7\%$$

$$\text{Case II} \quad \frac{5{,}500{,}000}{5} \times \frac{100}{3{,}000{,}000} = 36.7\%$$

$$\text{Case III} \quad \frac{7{,}000{,}000}{5} \times \frac{100}{3{,}000{,}000} = 46.7\%$$

The third method, which we shall describe in some detail, has come very much to the fore over the past twenty years or so. It involves very much more calculation work than the previous methods and this may have partly accounted for it being little used until comparatively recently when the advent of electronic calculators radically reduced the work of calculation.

This method can be applied in two ways but the basis of the calculations is the same, namely, the use of Discounted Cash Flows (DCF). The important advantage that this gives is that the timing of the cash flows are taken into account. The importance of this is quite simply that money received today by an investor (individual or company) is worth more than the same sum of money received at some time in the future. Money received today can be invested to earn interest whereas money received at some later date does not earn interest during the intervening period.

Suppose an investor can obtain 10 per cent per annum on an investment. The \$1 today becomes \$1.10 after one year, \$1.21 after two years, \$1.331 after three years, and so on. To such an investor it makes no difference whether he receives \$1 today or \$1.10 in one years time or \$1.21 in two years time, etc. We can also make this comparison the other way round — that is since \$1.10 in one years time is worth \$1 today we can equally say \$1 in one years time is worth:

$$\$\frac{1}{1.10} = \$0.9091 \quad \text{today}$$

This is known as the Present Value of \$1 in one years time at 10 per cent per annum, compounded annually. Similarly the present value of \$1 in two years time for the same conditions is $\$(1 \div 1.21) = \$0.8264$, and clearly the further away in time that the \$1 is received, so the lower is its present value. The lower the rate of interest, the higher the present value and vice versa.

Present values can thus be easily calculated for any particular rate of interest and books of tables are available showing present values over wide ranges of rates of interest and periods. Tables are also available showing the present value of unit income per period (e.g. per annum) for different numbers of periods and different rates of interest. Such tables can save work where steady cash flows are involved. Modern electronic calculators can however perform the calculations without the necessity to refer to tables, but it is important nevertheless to fully understand the methods of calculation and the reader is advised to perform a few calculations using tables to ensure a full understanding of the principles.

As already mentioned, there are two approaches to the analysis using DCF methods — the first is to sum the present values of the cash flows to find the Net Present Value (NPV) assuming a particular rate of interest, while the alternative is to find what rate of interest is required to give a net present value of nil, that is to find the rate of interest earned on the money invested in the project from time to time. We consider first the NPV method.

To use the NPV method we must first choose a rate of interest or discount rate. This is normally the minimum rate required by the investor and can be the rate of interest obtainable elsewhere, or being currently earned by the business, or expected by the shareholders — the choice will be further discussed later in this chapter. Suppose we choose a rate of 10 per cent and use this to calculate the NPV of our project:—

Table 8

Year	Net cash flow	P.V. factor (10%)	Present Value
0	(3,000,000)	1,0000	(3,000,000)
1	1,000,000	0.9091	909,100
2	1,000,000	0.8264	826,400
3	1,000,000	0.7513	751,300
4	1,000,000	0.6830	683,000
5	1,500,000	0.6209	931,350
	2,500,000	Net Present Value	1,101,150

Thus the project has a positive NPV of $1,101,150 showing that it is now worth this sum more than an investment at 10 per cent assuming the risks to be the same. If 10 per cent represents the interest generally obtainable on investments with similar risks attached, then if the opportunity to invest in the project was sold to another company, the sale price should be $1,101,150. If there are greater risks attached to the project then this would be a maximum figure.

We can also find the NPV's of Cases II and III of our project:

Table 9

Year	P.V. factor (10%)	Net Cash Flow	Present Value	Net Cash Flow	Present Value
0	1,0000	(3,000,000)	(3,000,000)	(3,000,000)	(3,000,000)
1	0.9091	250,000	227,275	1,000,000	909,100
2	0.8264	750,000	619,800	1,000,000	826,400
3	0.7513	2,000,000	1,502,600	1,000,000	751,300
4	0.6830	1,000,000	683,000	2,000,000	1,366,000
5	0.6209	1,500,000	931,350	2,500,000	1,552,250
		2,500,000	NPV=964,025	4,500,000	NPV=2,405,050

Comparison of Cases I and II shows how the change in the timing of receipts during the first three years is reflected in a change in the NPV which thus differentiates between Cases I and II, unlike the earlier methods of appraisal that have been described. Case III is shown to be the best project as one would expect. All three cases show a positive NPV and so if the discount rate used (i.e. 10%) represents the minimum acceptable to the investor, then each of the projects would be acceptable.

One point about the use of the NPV method must be made — if it is used to compare projects with differing initial investments, the NPV's will clearly depend in part on the size of the investment — if earning the same rate of interest, they would be proportioned. Ranking can be therefore achieved by dividing the NPV of the cash inflows by the NPV of the cash outflows to obtain a profitability index.

Thus for Case I we have:

$$\text{Profitability Index} = \frac{\text{NPV of cash inflows}}{\text{NPV of cash outflows}} = \frac{4{,}101{,}150}{3{,}000{,}000} = 1.367$$

and for Case II

$$\text{Profitability Index} = \frac{3{,}964{,}175}{3{,}000{,}000} = 1.321$$

The alternative way in which DCF's can be used is to find the actual rate of interest on the money invested rather than determining whether or not this rate is above or below some standard figure. The actual rate of return earned on the money actually invested in the project from time to time is referred to as the Internal Rate of Return (IRR) and the method is usually known as the IRR method. Unfortunately different writers on the subject have used different names and have referred to the method as the 'yield' method, the DCF method, the rate of return etc. The advantage of this method over the Net Present Value method is that it provides a more readily appreciated measure of the margin available in the project over the minimum return required or over a risk-free investment, the margin then being compared with the amount of risk attached to the particular project. Thus if the IRR for a particular project is found to be 15 per cent and the return on a risk-free investment is 10 per cent the margin to cover the risks of the project is readily seen, expressed as a return on capital, which the basis on which investors are most used to reviewing investments. The other advantage is that where a company wishes to use its cost of capital as the minimum acceptable rate, the problem of determining precisely what this rate is does not affect the DCF calculation whereas it must be used as the basis of the NPV calculation.

The IRR can be defined as that discount rate which equates the present value of the outward net cash flows with the present value of the inward cash flows. It is thus the discount rate that will produce an NPV of zero and so using PV tables we can find this rate by trial and error. This somewhat tedious process can be avoided by using one of many pocket electronic calculators which include IRR calculations in their capability. Nevertheless we shall demonstrate the principles by considering our project — Case I, where we have already seen in (Table 8) that a 10 per cent discount rate produces a substantial positive NPV.

Table 10

Year	Net cash flow	P.V. factor (20%)	P.V.	P.V. factor (23%)	P.V.
0	(3,000,000)	1,0000	(3,000,000)	1,0000	(3,000,000)
1	1,000,000	0.8333	833,300	.8130	813,000
2	1,000,000	0.6944	694,400	.6610	661,000
3	1,000,000	0.5787	578,700	.5374	537,400
4	1,000,000	0.4823	482,300	.4369	436,900
5	1,500,000	0.4019	602,850	.3552	532,800
	2,500,000		191,550		(18,900)

By interpolation IRR = 22. 73 per cent.

Interpolation will give a reasonably accurate answer over a small range — the actual IRR in this case is 22.71615 per cent.

It is always possible to check the calculated IRR by calculating and compounding the interest on the capital outstanding during each period:—

Table 11

Year	Capital Outstanding at start of period	Interest at 22.71615%	Net cash flow	Capital o/s at end of period
0	nil	nil	(3,000,000)	3,000,000
1	3,000,000	(681,485)	1,000,000	2,681,485
2	2,681,485	(609,130)	1,000,000	2,290,615
3	2,290,615	(520,339)	1,000,000	1,810,954
4	1,810,954	(411,379)	1,000,000	1,222,333
5	1,222,333	(277,667)	1,500,000	nil

This table illustrated clearly how the IRR is the interest earned on the capital invested in the project from time to time.

To further illustrate the relationship between NPV and IRR we can plot the NPV for different discount rates for Case I since Tables 8 and 10 give us four points. The curve cuts the x-axis at the IRR i.e. at zero NPV, and the y-axis at the cumulative net cash flow i.e. at zero IRR.

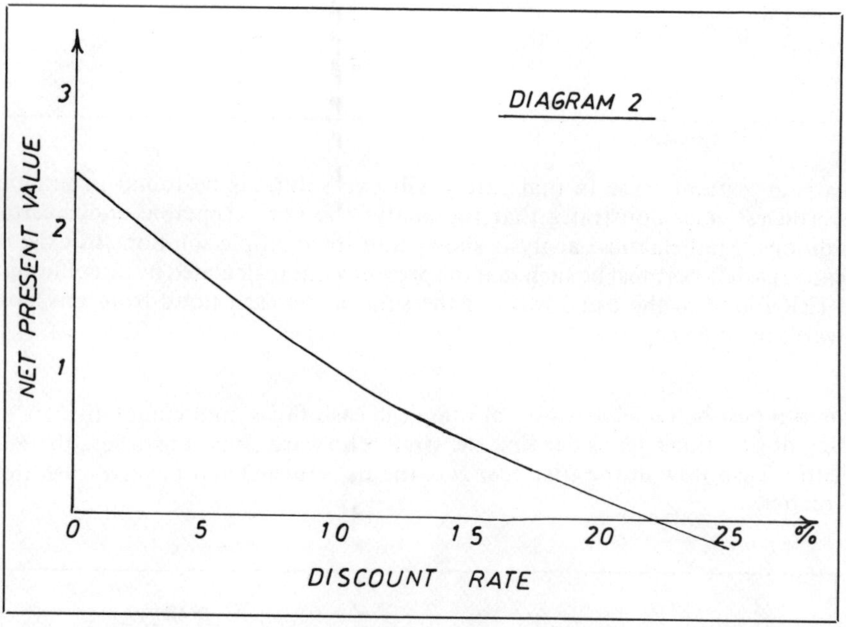

There is one situation where the IRR method can give an incorrect answer; this is the situation where there are negative cash flows at a late stage in the project. This is unusual in shipping projects where the investment (and therefore negative cash flows) is usually all at the beginning. However consider the following simple project:

Table 12

Year	Net Cash Flow
0	(1,000)
1	2,230
2	(1,240)

If we calculate the IRR carefully we will find that in fact there are two. We can illustrate this by plotting NPV's against discount rate.

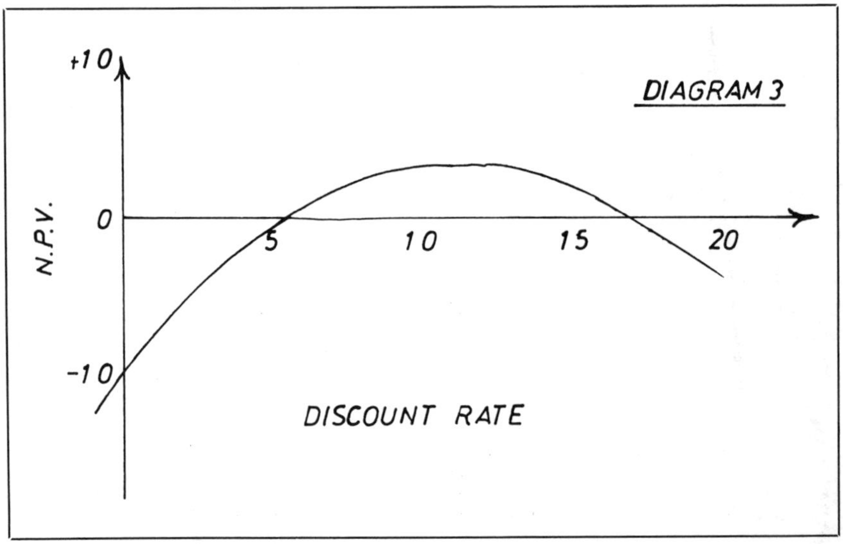

This is an extreme case in that rarely will two solutions be found in practice. Nevertheless it demonstrates that the analysis is not acceptable under certain conditions. Mathematical analysis shows that for multiple solutions to exist the negative cash flows must be such that the present value (calculated by discounting at the IRR found in the usual way) of the sum of the cash flows from any point onwards must be negative.

This can best be understood by plotting the cash flows, and cumulative PV's in different situations. Consider first the straightforward situations where there is a negative cash flow in the first year (i.e. the investment) and positive cash flows thereafter.

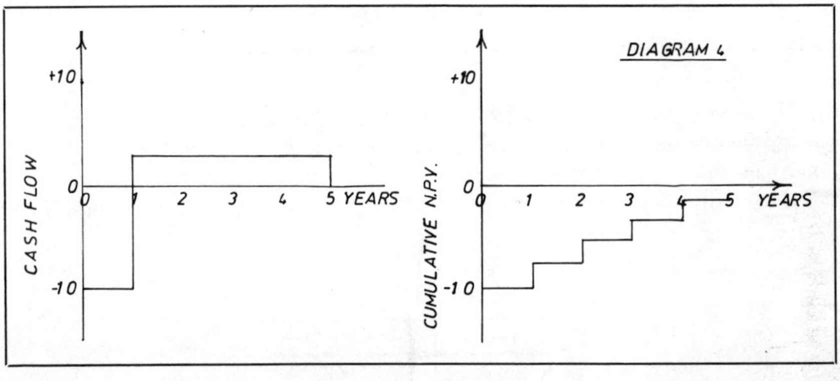

Now consider the case with a negative cash flow in the last period.

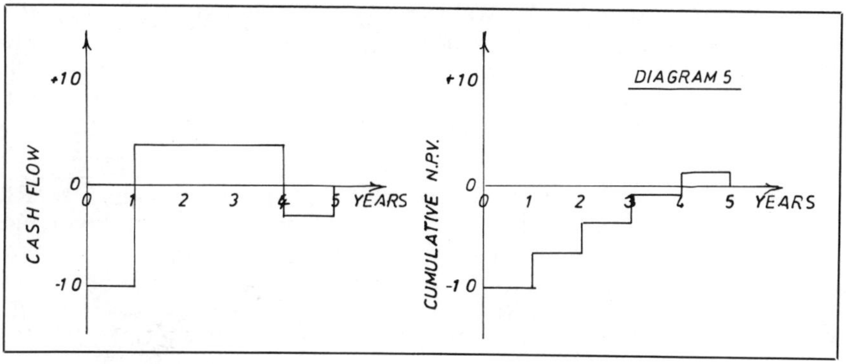

The problem derives from the fact that there is capital in hand as shown in the second diagram above which discounting assumes can be invested at the IRR. If there is such capital in hand it follows that the cumulative present values must lie above the base line at this point and that therefore the sum of the PV's of the cash flow from this point onwards must be negative.

Examination of the cash flows in the project in Table 12 shows that in the last (the second) year of the project there is a negative cash flow. Some of the cash received in the previous year would therefore have to be retained to meet this expenditure. However in discounting the negative cash flow at the IRR we have assumed that the money retained can be reinvested at this same IRR which, especially if the IRR is above the minimum acceptable, is not likely to be the case. It would be logical to assume either that the money retained was invested in a convenient risk-free investment until required or, as some analysts suggest that it is invested at the same rate as the company's cost of capital. If we assume that this rate was 5 per cent in the project in Table 12 then the sum to be retained from the cash flow in year 1 would be \$(1240 ÷ 1.05) i.e. \$1181, leaving a net cash flow in year 1 of \$1069.

The IRR would then be calculated using this cash flow for year 1 and nil for year 2, giving IRR = 6.9 per cent.

In practice it will usually be obvious whether or not this procedure should be followed, but if in doubt the test already described can be applied as a check.

Table 13

Year	NCF Project A	NCF Project B	NCF Project B — Project A
0	(2,000)	(3,000)	(1,000)
1	669	959	290
2	669	959	290
3	669	959	290
4	669	959	290
5	669	959	290
	1,345	1,795	450
IRR	20%	18%	13.8%

So far we have considered NPV's and IRR's primarily as a means of comparing a given project against a selected yardstick e.g. the company's cost of capital. Many situations arise where comparison of one project with another is required on a mutually exclusive basis i.e. only one of the projects can proceed. It may in fact be different ways of servicing the same requirement.

In such circumstances it does not follow that the project with the highest IRR is necessarily the best alternative open to the investor. Consider two mutually exclusive projects, A and B, as shown on page 43.

Project A with an IRR of 20 per cent would appear preferable to Project B with an IRR of 18 per cent. However in the third column of the table the difference in the cash flows of the two projects has been considered and it is seen that the additional investment required for Project B by comparison with project A produces in itself an IRR of 13.8 per cent. If the minimum return required is less than this then, in theory at any rate, the company should proceed with Project B. In practice however the directors may well consider that the additional risks involved in Project B are greater than the profit margin indicated by an IRR of 13.8 per cent and therefore decide to proceed with Project A. This type of analysis is known as the 'incremental' method since it considers the incremental cash flows consequent upon the choice of one alternative in preference to the other. The method should be used for comparing a mutually exclusive alternative with another. It is of course dependent upon a minimum level of return having been set if a positive answer is required and of course both projects must themselves be considered acceptable in isolation.

Having considered the basis for both the NPV and IRR methods of project appraisal we can examine their use in the typical situations that arise in shipping.

There are two sets of circumstances where use of these methods of investment appraisal may play an important part in investment decisions. Firstly there is the case where a long term charter may be on offer, such that the net cash flows throughout the asset's life can be calculated with some degree of confidence, or where the directors feel confident that they can predict the vessel's earnings even without a charter for the full period — in such cases the requirement is to establish whether or not the profits justify the risk.

The second situation is where the directors are confident that there will be a requirement for a vessel but are unable to predict the earnings year by year — the type of situation that often exists when vessels are ordered on spec for the tramp market. In this case the analyst has to compare different offers from shipyards, perhaps including second hand options.

Let us consider a typical project of the first type to illustrate the IRR method in a shipping context. A vessel is to be constructed over a two year period and will be employed from delivery from the shipyard on a fifteen year time charter. The vessel costs £10m, and will be financed by an 80 per cent loan repayable over $8\frac{1}{2}$ years with interest at $7\frac{1}{2}$ per cent pa. The payments during construction will be as set out in Tables 1-3; there is no interest roll up. There are fees etc., payable on signing of £100,000 and costs additional to the contract of £200,000 payable at delivery. The time charter will produce £1,650,000 pa after all operating expenses etc. The vessel will be sold for £500,000 at the end of the charter. Inflation is to be ignored — methods of dealing with it will be found in Chapter 10.

For a preliminary review of this project we will work on annual cash flows, but for a more detailed study six monthly periods may be preferred. The annual loan repayments shown are the sum of the semi-annual payments shown in Table 4. We shall also make our analysis by expressing all amounts as a percentage of the contract price.

Table 14

Year	Capital expenditure on contract	Other non-operating expenditure and income	Drawdowns	Loan Capital repayments	Loan Interest payments	Net charter income	Net cash flow
(1)		(2)		(3)			(4)
0	(5.00)	(1.00)	—	—	—	—	(6.00)
1	(40.00)	—	35.00	—	—	—	(5.00)
2	(55.00)	(2.00)	45.00	—	(4.32)	—	(16.32)
3	—	—	—	(9.41)	(5.82)	16.50	1.27
4	—	—	—	(9.41)	(5.12)	16.50	1.97
5	—	—	—	(9.41)	(4.41)	16.50	2.68
6	—	—	—	(9.41)	(3.71)	16.50	3.38
7	—	—	—	(9.42)	(3.00)	16.50	4.08
8	—	—	—	(9.41)	(2.29)	16.50	4.80
9	—	—	—	(9.41)	(1.59)	16.50	5.50
10	—	—	—	(9.41)	(0.88)	16.50	6.21
11	—	—	—	(4.71)	(0.18)	16.50	11.61
12	—	—	—	—	—	16.50	16.50
13	—	—	—	—	—	16.50	16.50
14	—	—	—	—	—	16.50	16.50
15	—	—	—	—	—	16.50	16.50
16	—	—	—	—	—	16.50	16.50
17	5.00		—	—	—	16.50	21.50
	(95.00)	(3.00)	80.00	(80.00)	(31.32)	247.50	118.18

Notes: (1) It is sometimes preferred to label the year of delivery of the vessel year 0 the preceding years being labelled -1, -2, etc.
(2) This expenditure may well be capitalised in the owners books but that is immaterial for cash flow purposes.
(3) Actually 9.41176 rounded off to 9.41 (9.42 in year 7 to give correct total repayment).
(4) Totalling column and comparing with the totals of the preceeding columns provides a useful arithmetic check.

To obtain the IRR one would then proceed by trial and error as already described (Table 10) or by the use of a suitable calculator. (IRR = 17.273 per cent)

A calculation such as this rarely occurs in isolation, in that often the contract price of the ship changes and this is one reason for working in percentages of the contract price — furthermore once the capital and interest payments for any given loan have been calculated they can be applied to any value of vessel without further calculation. With the possibility of the charter income also changing it may be considered worthwhile to calculate the net cash flow without the charter income i.e. the cash

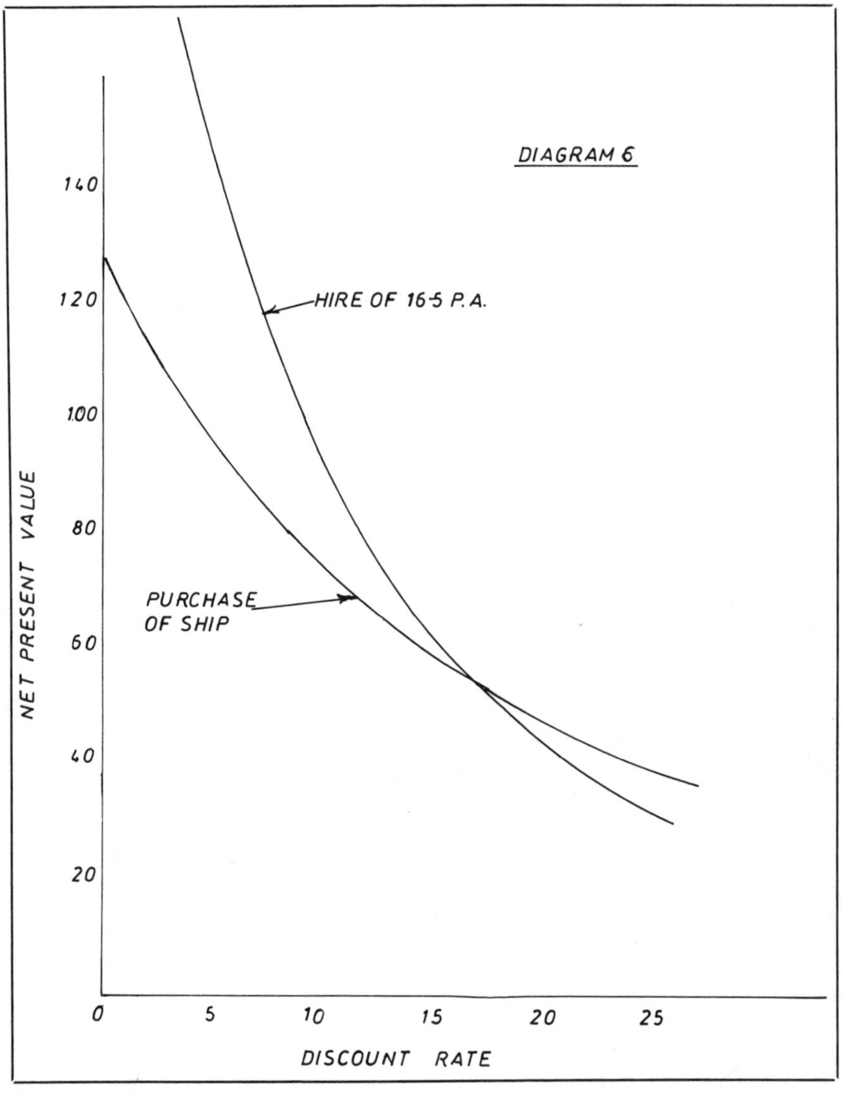

flow related to the acquisition of the ship. The present value of the cash flow can then be calculated for a range of discount rates and plotted as in Diagram 6. The present value can then be calculated separately for any given net income or pattern of net income and plotted on the same axis to obtain the IRR graphically as indicated in the Diagram.

It is worth remembering that the NPV of a cash flow is the sum of the NPV's of its components and it is quite in order to find the NPV of different parts of an overall cash flow and sum these NPV's as an alternative to summing the cash flows in each period and then finding the total NPV. The important point to watch is that the same discount rate is used in each case.

In practice is is usually the IRR after tax which will be required; the way in which taxation is taken into account in these calculations is given in Chapter 8; the problem of inflation is covered in Chapter 10.

When the IRR has been calculated it can then be compared with the minimum rate required in the light of the risks attached to the project and other factors that the directors may consider relevant. This brings us back to the question of what should be used as the minimum rate.

Analysts usually recommend that the minimum rate should be the company's cost of capital, that is the average overall percentage return that the company pays (or that its shareholders expect it to pay) on its various sources of capital — clearly no company would wish to make any long term investments at a lower rate as it would be better to repay the more expensive forms of capital if no suitable projects could be found. However it is usually difficult to calculate the cost of capital for any well established company and so other yardsticks must be considered. One that may in any case be considered more appropriate in some circumstances (e.g. a company with surplus cash) is that which the company would obtain in a risk-free investment such as gilts. It is interesting to note that surveys of British industry have shown that the most commonly used yardstick is the company's overdraft rate!

In deciding the minimum acceptable rate we are moving from the area of financial analysis into that of business decisions and the choice of rate is usually made by the directors. What is clearly important is that the financial analyst then presents the directors with information which enables them to compare the profitability of any project, as assessed by the analyst, with their assessment of the relative risks attached to the project. It is therefore worth considering some basic principles of investment.

When money is lent or invested the lender can expect first some compensation for giving up the opportunity for immediate use of funds, and for loss of value due to inflation during the period of the loan. If there is no risk attached to the loan or investment this compensation can be equated to the return on a 'safe' investment, e.g. gilt-edged stock.

However is there a risk that some or all of the money will not be returned to the lender then additional compensation is required to cover such a risk and clearly the greater the risk the greater should be the compensation, i.e. the profit, if the venture being financed is successful.

Thus when considering the level of return required to justify a new investment it can be divided into two parts:—

1. The return for giving up the use of the money for the period (the gilt-edged return).
2. The profit, i.e. the reward for the taking of the risk of losing some or all of the capital invested.

If there is no risk in the venture the return on the total capital invested should at least equate to that of gilt-edged investments having a similar life (in a given venture this will probably be a mix of different maturities). When there is a measure of risk in the venture this should be reflected in the difference between the total return earned from the venture and this gilt-edged return.

Thus when considering new ventures, and when comparing alternative ventures, one yardstick should be this difference between the rate of return calculated on the total capital invested (i.e. at risk) and the redemption yield of the alternative investment in gilt-edged securities.

Ideally therefore one should ensure that the projected return on a new project is both above the company's cost of capital and gives a sufficient margin over a risk-free investment to justify the risks of the project.

There is also another factor which should rightly be taken into account in determining margins etc. The DCF basis of calculation assumes that capital can be drawndown and repaid as and when required without penalty. In practice this is hardly likely to be the case (unless marginal financing is by overdrafts) and there is likely to be cash sums held perhaps on short term deposit. Such cash will earn somewhat less than the IRR (or else the project would not be justified) and so if taken into account as being necessary to the project the existence of such sums would in fact lower the overall return on the investment; this situation must therefore be taken into account as one of the factors determining the minimum acceptable return even though its effect is small.

Having considered the case of an isolated project where the directors are required to make a simple 'go' or 'no-go' decision, we now look at the situation where there is a choice of possibilities which are mutually exclusive — that is where the decision has to be made between alternatives as well as deciding whether or not to proceed with any of them.

Using the IRR method the IRR's for the alternatives may give an immediate preferred choice but if more than one gives an IRR greater than the acceptable minimum an incremental analysis (as described on p.43) should also be carried out.

A comparison that must frequently be made is that between alternative shipbuilding offers. Where the alternative vessels will produce identical incomes from trading (i.e. their operating costs etc., will be identical) the initial analysis is often simplified by comparing the NPV's of the costs shown in the offers — indeed where the vessel is being ordered 'on spec' this is often the full extent of the analysis. (We should strictly compare profitability indexes since there are differing amounts of capital invested by the owner, but at any given discount rate the NPV of the cash inflows will be the same; the 'best' project must therefore be the one with the lowest NPV of the cash outflows.) An important point to realise in this case is that the relative merits of the alternatives may change as the discount rate chosen changes.

Table 15

Cost		Financing Arrangements
Vessel A	100.00	80% loan over 8½ yrs at 7½% pa + 10% loan over 8½ yrs at 14% pa
B	100.00	80% loan over 8½ yrs at 7½% pa only
C	85.00	30% loan over 6 yrs at 14% pa only (30% of 100)

Consider the case where a vessel is to be ordered from one of the following alternatives where each vessel is technically similar with identical operating costs, etc, each will produce the same annual income for capital servicing and have the same residual value. The alternatives can then be compared on the basis of their purchase costs only.

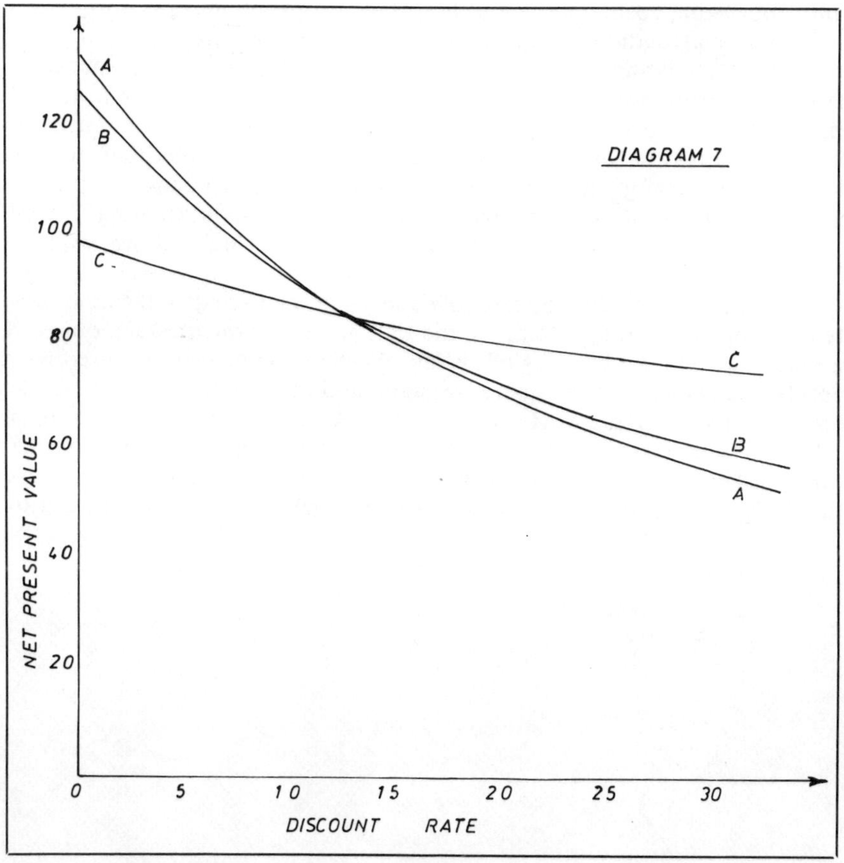

If the NPV's of the three alternatives are calculated assuming they are paid for at delivery with the owner providing the non-loan finance, the following diagram shows how the relative merits of the alternatives will depend very much on the discount rate considered appropriate.

We shall consider an alternative approach in appendix 1 but first consider another matter to conclude this chapter.

In all the DCF analysis considered we have divided the life of the project into a number of equal periods (e.g. six months) and then made the assumption that the net cash flow for that period actually occurs at the end of the period. Such an assumption is necessary for practical reasons, but it is prudent to check the pattern of the cash flows within each period — in the case of new construction this will probably be done in any case for the building period to evaluate the various costs and expenses. During the service period however we have to ensure that we have not overlooked a possible requirement for working capital as a result of our assumption.

Suppose for example the vessel's operating costs were all paid six monthly in advance and the hire six monthly in arrears. If our analysis was based on six month periods then in this extreme case the funds invested in the project from delivery of the vessel until the end of the project would in fact be greater by the cost of six months operating costs than would have been assumed; and therefore not only would additional funding be required but the rate of return on the investment would be less than that shown by the analysis. The effect of an extreme situation on the analysis in Table 14 would be to reduce the IRR from 17.273 per cent to 13.957 per cent if running costs of £1,100,000 per annum (11 per cent of the contract price) were paid annually at the beginning of each annual period rather than at the end. This is of course a far greater difference than would occur in practice but it serves to illustrate that the point cannot be overlooked — apart from the lowering of the IRR, the owner is required to invest additional cash — £1.1m in the above case).

In practical cases the effect of carefully analysing the timing of the cash flows is often the opposite and is to improve the IRR and lower the capital required. For example, a vessel built for long term time charter from delivery may receive hire monthly in advance. Yet the loan repayment and interest payments are made six monthly in arrears; wages will be paid monthly in arrears, repairs and maintenance will be paid well in arrears, and the only significant item paid in advance may be the insurance. Thus in such a case there will generally be cash in hand which however is generally ignored for purposes of analysis being regarded as of short term duration.

Chapter 7

Using Foreign Currency

Few companies involved in any aspect of shipping can fail to be involved in foreign exchange deals at some time or other. In this chapter we are concerned primarily with the problems and opportunities offered when shipbuilding contracts are in a foreign currency, and start with some explanations regarding the foreign exchange market.

Whenever a business deal is concluded between parties in different countries, one or other of the parties must be involved in a foreign exchange deal. For example if a shipowner in the UK sells a ship to an owner in France, either the French owner will ask his bank to sell to him the necessary amount of sterling with which he can pay the UK owner, or the UK owner will accept French Francs in payment which he will then sell to his bank for Sterling.

When an exchange of currencies (i.e. the sale and purchase) is agreed the date for the actual delivery of the currencies is also agreed. In 'ordinary spot' deals the delivery of the currencies (which must both take place on the same day) takes place two working days after the deal is arranged. However the delivery can be arranged to take place on any mutually convenient date in the future — deals in which delivery takes place within a week are known as 'spot' deals; those over a week ahead are 'forward' deals.

Both the 'spot' and 'forward' rates at which currency is bought and sold vary from day to day and indeed from hour to hour. The forward rates are linked to, and vary with the spot rates. Many factors influence the supply and demand situation which determines the variation of spot rates.

However in shipping it is the forward rates which are usually important, particularly in a building contract for example, and so we now consider the principal details of forward deals.

A contract for the sale and purchase of a currency can be made with the delivery of the currencies on an agreed date as far ahead in the future as the buyer and seller wish. In practice the ease with which buyers and sellers can be found for forward contracts depends primarily upon the amount of trading in the two currencies — for example if Sterling and US Dollars are the currencies concerned, forward contracts can be arranged several years ahead while in some of the lesser traded currencies a few months ahead may pose problems.

Forward rates are always quoted relative to the ordinary spot rate. If the currency in question is more expensive for forward delivery than for spot delivery it is said to be at a premium, whereas if it is less expensive it is said to be at a discount. For example:—

Spot US Dollars for Sterling :	\$1.5155 (per Pound Sterling)
Three months forward	0.70 cents premium
Therefore Three months forward rate	=\$1.5085

Note that US Dollars are quoted at a premium in this example which means they are more expensive if purchased forward, i.e. one gets fewer Dollars per Pound. If they were quoted at a discount then one would get more Dollars per Pound for a forward purchase. Conversely to an American Sterling would be at a discount in the example shown above.

The leading financial papers, such as the British Financial Times, carry daily details of the previous days movements of the ordinary spot rate, the closing rate (i.e. the ordinary spot rate at the close of business) and selected forward rates. Some papers also give 'cross rates' which are the rates of exchange between two 'foreign' currencies — for example the US Dollar for French Francs would be a cross rate to a UK resident.

In shipping there is often a choice between entering into a forward contract as soon as a future requirement for foreign currency becomes known, ('covering forward') or waiting until the times comes and dealing spot which amounts to speculating on currency rate movements. Before making any choice between these alternatives it is necessary to appreciate the factors determining forward rates.

Suppose that you have a sum of money to invest for one year. You could invest it in a risk free government stock and earn interest at say 10 per cent per annum. However you notice that in another country equally risk free government stock is paying 15 per cent per annum, so you ask your bank to buy the foreign currency with your sum of money so that you can invest in the alternative stock. Knowing the amount of foreign currency that you will have in a years time you ask your bank to also arrange to convert this back into your own currency one year hence. Clearly if the forward rate was the same as the spot rate you would be 5 per cent better off (less brokerage!) as a result of this manoeuvre than if you had remained in your own currency throughout. Not surprisingly however others will have noticed this difference before you with the result that demand will have created a discount on the forward rate of about 5 per cent and because of this you will be no better off than if you invested in your own country at a lower rate!

This type of manoeuvre is known as interest arbitrage. The theory of interest arbitrage is that the difference between the spot and forward exchange rates will equal the difference in interest rates between similar types of investments in different countries (and currencies).

In practice this theory holds good while the situation is stable, however when interest rates do change large amounts of money may move until the position stabilises at the new rates. Such movements are not acceptable to many countries and are one of the reasons that countries impose Exchange Controls which will in fact contribute to the time lag in re-establishing equilibrium by restricting the flow of funds.

Perhaps the most important consequence of the fact that forward rates are determined by differences in interest rates is that forward rates do not therefore necessarily reflect in any way the market's view of future changes in the spot rate.

When a shipowner for example is faced with the choice of covering forward or not, there are many who will advise him to do so rather than to 'speculate', perhaps on the grounds that his business is shipping and not foreign exchange. On the other hand it can be argued that dealing in foreign currency is all a part of shipowning and that in any case there are times when the costs of covering forward bear no relation to the risks being avoided. Because a shipowner is inevitably involved in foreign exchange transactions it would seem sensible for him to keep in close touch with the foreign exchange markets in order that when a forward requirement arises he is in a position to assess the matter himself even though the advice of his bankers would probably be obtained as well.

We can now consider the effect of the different currencies on the choice between a number of alternative building contracts with loans in different currencies. In some cases there may be a choice of currencies with a particular building contract. Bearing in mind that on signing a building contract with an eighteen month building period and a eight and a half year loan after delivery the owner will have debts of up to ten years ahead, the question of the currency in which these debts are incurred can be vital.

Given a free choice, what factors should determine the choice of currency? It is often suggested the loan should be in the same currency as that in which the hire will be paid, i.e. the currencies will be matched and the currency risk eliminated. Thus if charter hire is in US Dollars then the loan should be in US Dollars. However this can be a false argument.

Consider the case of a vessel chartered for short periods at the market rate. This rate will be determined by the supply and demand for suitable vessels in the short term; on the long term it will reflect building costs and operating costs in a number of countries and currencies. Thus although the US Dollar may be used to express the charter rates the actual rates will reflect the value of a number of currencies and in this situation there is no 'matching' currency. The vessels earnings will have a certain value at the time they are paid which will be available to service the loan. Since we are unable to change this value (being determined by the market) our main concern in the choice of currency for the loan should be that the value of the loan repayments are at a minimum — i.e. the loan must be in a weak currency.

A different situation arises where a vessel is employed on a long term charter where the hire to be paid is specified in a certain currency — in this case a 'matching' loan will eliminate the currency risk (in so far as the loan repayments are concerned).

The important point to note is that it is the currency in which the value of the vessels earnings will be established that is important — this may not necessarily be the currency in which the earnings will be paid.

It should be noted that the OECD agreements referred to in Chapter 3 fix the rate of interest at the same level in all currencies, with the result that the amount of subsidy may vary considerably in different countries.

How then can we take different currencies into account when comparing alternative contracts? There are two obvious possibilities: one is to work on the basis of forward rates — appropriate if the intention was to buy forward — the other to work on the market's view of forward rates which would be the alternative to use if no forward cover was intended. There is of course a third alternative, that if ignoring the possibility of future currency movements — an unrealistic assumption but one which might be justified if there were conflicting views in the market.

When forward cover can be obtained at a discount this in effect lowers the price payable for the vessel — when forward cover is at a premium the price is raised. The elimination of the currency risks by forward cover can also make an appreciable difference to the cash flows and thus to the results of the comparison of bids from different countries. In some countries a choice of currencies for the contract and loan may be possible and such alternatives are always worth considering by an owner.

There may be occasions when doing investment appraisal that it is more convenient to do the calculations in a foreign currency and then convert the end result rather than convert the inputs into one's own currency before performing the calculation. Such an approach however needs considerable care.

The position during the building period in the case of contracts in foreign yards must also be considered in respect of the owners cash payments, and consideration given to the forward purchase of the necessary currency. Indeed the possibility of substantial changes in exchange rates has sometimes influenced owners in their negotiation of the stage payments — for example in 1971 when it was thought that revaluation of the Japanese Yen was likely, foreign owners negotiating building contracts based on the Yen were in some cases arranging a 50 per cent signing instalment!

Where forward cover is required there are three possible difficulties. Firstly there may be problems in covering very far ahead and certainly there will be difficulty in covering for say ten years ahead even in the leading currencies. The advice of a bank should be sought — deals well ahead can sometimes be arranged if a few days are allowed for a suitable exchange contract to be arranged. The second problem that can arise is that the bank arranging the forward contract will need to be satisfied of the owners financial ability to meet the contract, because if rates move against him he will be buying currency at well above the market rate.

The third difficulty concerns the timing of the forward contracts. During the building period the instalments are usually based on the progress of construction work (p.15) and the building yard will probably only be able to give an estimate of the likely dates upon which the payments will fall due. Yards building for the offshore industry usually plan construction in far more detail than a typical shipyard and are usually more reliable in their estimate of construction progress. Furthermore in the event of a change in delivery date all the loan repayment dates will change and the owner will be buying foreign currency perhaps long before he requires it. There is however a solution to this problem in the form of the 'swap', which is a pair of foreign exchange deals in the same currencies but in different directions on different dates. Thus an owner having completed his forward exchange contract on the prearranged date enters into a swap whereby the foreign currency is sold immediately and repurchased at the new date at the forward rate then ruling. Thus no new risks are incurred — the cost being the difference between the spot and forward rates as on the day the initial purchase of the pair is made.

There is an alternative to a swap where the owner anticipates from the beginning that the date on which he is actually going to need the foreign currency may change — this is the 'option' forward contract.

In such a contract the owner arranges the forward contract with his bank at a fixed rate of exchange but for delivery on a date to be decided by him later. The owner has to pay for this flexibility and so generally the swap alternative is to be preferred, even though only the option gives complete protection against currency risk.

One method of reducing currency risks in connection with loans is to make the loan, not in a single currency, but in a combination or "basket" of different currencies. Although this does reduce the possibility of extreme fluctuations between the loan and the currency which determines the value of the borrowers' income, it does not help if the latter currency moves relative to other currencies. This method has not therefore proved popular.

No chapter on the use of foreign currencies would be complete without mention of Euro-currencies, and in particular Eurodollars, the original Eurocurrency. What are Eurodollars? They are dollars in the bank accounts in New York belonging to central banks, industrial companies and others resident outside the USA, which are lent to other banks and companies also outside the USA. They originally came into existence because of restrictions on interest rates within the USA. Similarly Eurosterling is sterling held in accounts in London belonging to overseas residents, lent to other overseas residents etc. A number of factors have led to the growth of these Eurocurrencies ("Euro" only because their trading is centred on Europe, particularly in London) which now provide a major source of funds for commercial shipping loans. There is also the Eurobond market on which long term capital can be raised as a further possible source of funds.

Finally in this chapter we consider exchange controls whereby in many countries the government imposes restrictions on transactions involving the sale and purchase of foreign currencies.

In some countries any holder of domestic currency can buy foreign currency at any time, in which case the domestic currency is described as being 'fully convertible'. For such a state to exist the government must be satisfied that all such conversions will be covered by the rate of inflow of foreign currencies such that the country's holding of such currencies (its foreign exchange reserves) will still grow rather than diminish. Few governments can be this confident with the result that most countries have some form of exchange control.

In any contract involving a foreigner it is therefore prudent to check that the transaction does not fall foul of any such controls. Banks active in the country concerned are usually the best source of such advice.

Chapter 8

Taxation

The methods whereby governments derive their income through taxation vary from country to country and therefore in this chapter we consider some general principles rather than any one country's tax system although references will be made to the UK tax system where appropriate by way of illustration.

Owners registered in some countries are able to trade their ships without the profits being taxed, their only payments to the government being a registration fee — such are the 'flag of convenience' countries.

Most countries base their tax charge on the profits on trading (i.e. as shown in the P & L account) but with various modifications. The most important of these usually concerns the depreciation charge. In Chapter 1 we stated that the depreciation of an asset such as a ship would normally be spread over the life of the ship e.g. 15 years. However there are different methods of calculating the appropriate amount of depreciation to be charged each year. The most commonly used method for accounting purposes is the 'straight-line' method already described.

An alternative method is the reducing balance method whereby the charge equals a certain percentage of the 'written-down' value of the asset (that is, the original cost less depreciation charged to date). Thus if an annual charge of 18 per cent was used the annual charges for depreciation in the case of a vessel costing 100 would be:

1st year	18% of 100 = 18.00	Written down value=			100–18	= 82.00
2nd year	18% of 82.00 = 14.76	"	"	"	= 82.00–14.76	= 67.24
3rd year	18% of 67.24 = 12.10	"	"	"	= 67.24–12.10	= 55.14
		etc				
14th year	18% of 7.58 = 1.36	"	"	"	= 7.58– 1.36	= 6.22
15th year	18% of 6.22 = 1.12	"	"	"	= 6.22– 1.12	= 5.10

A comparison of the annual allowances above with the straight-line allowance of 6.33 per cent pa (which would give the same written down value after 15 years) shows that the declining balance method gives much larger depreciation charges in the early years and much smaller charges in later years by comparison to the straight line method.

The depreciation allowances for tax purposes — or capital allowances as they are usually called — are of two main types, the initial or first year allowance, and the annual writing down allowance.

In most countries the allowances are arranged to give accelerated depreciation for tax purposes, provided the profits have been made to enable the allowances to be used. In some cases there are no initial allowances, the allowances are restricted to an annual writing down allowance, usually calculated on a reducing balance basis. An example of this will be the UK system from 1986 onwards where only an annual writing down allowance of 25 per cent will be allowed.

In other cases, as well as an annual writing down allowance, an initial or first year allowance is granted in the form of additional allowances that may be used in the first year — these allowances can be added to the sum upon which the annual writing down allowance is calculated if they are not used in the first year. Until 1984 the UK allowed 100 per cent first year allowances on second-hand ships — any part of these allowances not used then forming the basis for 25 per cent writing down allowances in subsequent years.

Normally the total allowances i.e. first year allowance plus the total available for the annual writing down allowances, equal the capital cost of the vessel although there have been cases where the total allowances have exceeded this. For example at one time in the UK allowances were increased to 140 per cent in recognition of the increased replacement costs due to inflation.

The most advantageous use of capital allowances is that of 'free depreciation' where the allowances available can be used in each year to the extent required by the owner, but any allowances not used in the first year can be carried over and used to whatever extent the owner wishes the next year, and so on. This is the type of depreciation enjoyed by UK shipowners.

With capital allowances available in these ways it will be readily appreciated that they may be used at a far faster rate than that at which the vessel is being depreciated in the Annual Accounts, so that at some stage in the vessel's life the capital allowances will be exhausted, yet there may still be many years depreciation charges to be made against the profits in the accounts. This difference in the rate of use of the capital allowances and the rate at which the vessel is depreciated means that the profits after tax will be high in the earlier years and low once the allowances have been used. In order to avoid this the accountants may open a "Tax Equalisation" Reserve into which a proportion of the profits in the earlier years are transferred, the amount representing the difference between the tax that would have been paid if the capital allowances had equalled the accounting depreciation, and the tax actually paid. In later years this reserve would be used to increase profits. Opinions are however divided among accountants as to the value of this manoeuvre and some companies, principally those with large allowances available and low profits, no longer make transfers to such a reserve.

When a vessel is sold, the difference between the initial cost and the sale price (the net cost) will be calculated and compared with the total capital allowances used. If the capital allowances used are less than the net cost, an additional allowance (balancing allowance) equal to the difference is immediately made available; if smaller, a tax charge (balancing charge) is levied on the difference.

An important feature of capital allowances is that apart from eliminating any taxable profits in the year in which they are used they may, under some regimes, be used to create a tax loss that can then be carried back to eliminate the taxable profits of the previous year and thus save the tax due thereon. For example, suppose a shipowner pays tax at 50 per cent on profits of £500,000 in a year. In year 2 he makes profits of £1,500,000 and purchases a new vessel for £2,000,000. A 100 per cent first year allowance of £2,000,000 is available to eliminate the profit of £1,500,000 in year 2 and to create a tax loss in this year of £500,000. This tax loss can be carried back to eliminate the tax liability in year 1 — if tax for this year has already been paid it will be repaid by the authorities. The extent to which this type of transaction is allowable depends of course upon the tax legislation of the country concerned. In the UK a tax loss created in this way in a company can be carried back three years, but a tax loss must be created in each year in turn working backwards i.e. one cannot create a tax loss three years back leaving profits in later years, hoping to eliminate them by a loss the next year carried back.

Capital allowances may be available against the instalments paid on a building contract and some tax advantages may therefore be gained immediately on signing the building contract. This is the case in the UK, the allowances can be claimed irrespective of whether the instalment has been paid by the owner from his own resources or from borrowed money.

If the owner is contemplating using suppliers credit as opposed to the usual buyer's credit he should check carefully his entitlement to capital allowances, since they may be available only as the payments are made over the full period of the credit facility e.g. 8½ years from delivery. Suppliers credit may also affect certain types of subsidy, in that the authorities may pay the subsidy only against actual payments of the capital cost. This was the case with UK Investment Grants (1966-70) and using suppliers credit meant that the grant was paid over nine or ten years rather than all being paid by delivery of the vessel.

It is also worth noting that capital allowances are normally based on the contract price of the vessel as paid during the building period. If for example a UK owner is building overses the capital allowances will be based on the Sterling value of the instalments paid at the time of payment, whether or not the payments are made with currency purchased at the time, or from a foreign currency loan. In the latter case the owner then has to take the risk that the rates of exchange may increase (or decrease) the value of the loan expressed in his own currency (in which his accounts are made up) after delivery. Such changes in the value of the loan are not reflected by changes in the capital allowances — in most countries the 'loss' or 'profit' arising each year because of the revaluation of foreign currency loans is ignored for tax purposes; the choice of currency for a loan as discussed in Chapter 7 can therefore be of vital importance.

In tax regimes where there are no initial allowances, the difference between capital expenditure and revenue expenditure becomes particularly important since the latter will be allowed in full against taxable profits in the year concerned, while the former will be spread over a number of years. In the UK the distinction is a matter of case law. The distinction depends upon whether the payment was a lump sum or whether there would be recurring payments; whether some asset with lasting or enduring qualities was to be obtained; and the manner in which what was obtained was to be used. One important point for shipowners is that while repairs are normally revenue, repairs to a newly acquired ship relate to her service before acquisition are regarded as capital expenditure.

Table 16

Year	Capital Expenditure	Other non-operating exp. & income	Loan drawdowns	Capital repayments	Interest repayments	Net charter income	Net cash flow before tax	Profits for tax [1]	Cum. capital allowances b/f	Capital allowances used	Capital allowances c/f	Tax at 50%	Net cash flow after tax
0	(5.00)	(1.00)	—	—	—	—	(6.00)	—	—	—	6.00	—	(6.00)
1	(40.00)	—	35.00	—	—	—	(5.00)	—	6.00	—	46.00	—	(5.00)
2	(55.00)	(2.00)	45.00	—	(4.32)	—	(16.32)	(4.32)	46.00	—	103.00	—	(16.32)
3	—	—	—	(9.41)	(5.82)	16.50	1.27	10.68	103.00	6.36	96.64	—	1.27
4	—	—	—	(9.41)	(5.12)	16.50	1.97	11.38	96.64	11.38	85.26	—	1.97
5	—	—	—	(9.41)	(4.41)	16.50	2.68	12.09	85.26	12.09	73.17	—	2.68
6	—	—	—	(9.41)	(3.71)	16.50	3.38	12.79	73.17	12.79	60.38	—	3.38
7	—	—	—	(9.42)	(3.00)	16.50	4.08	13.50	60.38	13.50	46.88	—	4.08
8	—	—	—	(9.41)	(2.29)	16.50	4.80	14.21	46.88	14.21	32.67	—	4.80
9	—	—	—	(9.41)	(1.59)	16.50	5.50	14.91	32.67	14.91	17.76	—	5.50
10	—	—	—	(9.41)	(0.88)	16.50	6.21	15.62	17.76	15.62	2.14	—	6.21
11	—	—	—	(4.71)	(0.18)	16.50	11.61	16.32	2.14	2.14	—	—	11.61
12	—	—	—	—	—	16.50	16.50	16.50	—	—	—	(7.09)[2]	9.41
13	—	—	—	—	—	16.50	16.50	16.50	—	—	—	(8.25)	8.25
14	—	—	—	—	—	16.50	16.50	16.50	—	—	—	(8.25)	8.25
15	—	—	—	—	—	16.50	16.50	16.50	—	—	—	(8.25)	8.25
16	—	—	—	—	—	16.50	16.50	16.50	—	—	—	(8.25)	8.25
17	5.00		—	—	—	16.50	21.50	21.50	—	—	—	(8.25)	13.25
18	—	—	—	—	—	—	—	—	—	—	—	(10.75)	(10.75)
	(95.00)	(3.00)	80.00	(80.00)	(31.32)	247.50	118.18	221.18	103.00	—	(59.09)	59.09[2]	59.09

[1] Actually col. 7 plus col. 6.

[2] $(16.32 - 2.14) \times 0.50$

[3] Cols. 13 and 14 totals will only be equal if all expenses are allowable and tax is 50%.

Interest payments on loans are usually allowed against taxable profits — they are in the UK — but only if the company is trading, which means that in a one-ship company, some trading must be undertaken during the building period, if interest on the loan drawdowns is to be allowed against profits.

Owners will clearly be more concerned with the financial results of a new project after the resulting tax liabilities have been met, rather than before. The appraisal methods described in Chapter 6 should therefore take taxation into account. This can easily be done in the case of the first two methods described and is not difficult when DCF methods are used. Consider the example of the new project used in Chapter 6 (page 44).

We will first consider the effects of taxation on the project in isolation from the owners other tax liabilities. The calculation will be based on the pre-1984 UK tax system with 100 per cent first year allowances and free depreciation, with tax at 50 per cent (it was actually 52 per cent). The interest payments on the loan are allowed against the tax liability. We assume that the tax liability in any year is paid the following year.

The IRR after tax is 13.22 per cent compared with 17.27 per cent already calculated (Table 14) when tax is ignored or not paid.

Two points are worth noting about the above calculation. Firstly the fall in IRR is perhaps much less than might at first be expected when half the profits are taken in tax. The reason for the comparatively small fall lies in the timing — note that tax is all paid towards the end of the project. The present value of 1 in 12 years time at 13 per cent is 0.2307, and in 18 years time is 0.1108, so that the amounts of tax paid are heavily discounted in the calculation.

The second point of interest concerns the negative cash flow in the final (18th) year. If this is included in the calculation without any special treatment the calculated IRR is 13.276 per cent. Reference to Chapter 6 (page 43) will show however that negative cash flows such as this at the end of a project require special treatment. If we assume a rate of interest receivable on invested cash of 5 per cent net of tax then the sum of 10.75 ÷ 1.05 = 10.24 must be taken out of the previous years cash flow to cover this payment. The cash flow in year 17 should therefore be 3.01 with nil cash flow in year 18. If the IRR is then calculated on this revised basis the rate will be found to be 13.224 per cent compared to 13.276 per cent. This situation is usual where tax payments are made in arrears and clearly there is little need for the special treatment in a preliminary work on a new project of this type since the effect on the calculated IRR is small.

We can also consider the situation where there are no first year allowances — full calculation in this case is shown in Table 17.

It is assumed in this calculation that the expenditure in column 3 is capitalised for tax purposes. The IRR after tax in this situation is 12.67 per cent. Thus in the case of the isolated project (i.e. where the owner does not have other taxable profits that can be reduced by the use of capital allowances) the difference in the IRR between the more advantageous use of capital allowances — 100 per cent first year allowance and free depreciation — and a 25 per cent writing down allowance with no initial allowances and without free depreciation is small (13.22 per cent v. 12.67 per cent) if the rate of tax remains unchanged.

Table 17

Year	Capital Expenditure	Other non-operating exp. & income	Loan drawdowns	Capital repayments	Interest repayments	Net charter income	Net cash flow before tax	Profits for tax	Cum. capital allowances b/f	Capital allowances used	Capital allowances c/f	Tax at 50%	Net cash flow after tax
0	(5.00)	(1.00)	—	—	—	—	(6.00)	—	—	—	6.00	—	(6.00)
1	(40.00)	—	35.00	—	—	—	(5.00)	—	6.00	—	46.00	—	(5.00)
2	(55.00)	(2.00)	45.00	—	(4.32)	—	(16.32)	(4.32)	46.00	—	103.00	—	(16.32)
3	—	—	—	(9.41)	(5.82)	16.50	1.27	10.68	103.00	6.36	96.64	—	1.27
4	—	—	—	(9.41)	(5.12)	16.50	1.97	11.38	96.64	11.38	85.26	—	1.97
5	—	—	—	(9.41)	(4.41)	16.50	2.68	12.09	85.26	12.09	73.17	—	2.68
6	—	—	—	(9.41)	(3.71)	16.50	3.38	12.79	73.17	12.79	60.38	—	3.38
7	—	—	—	(9.42)	(3.00)	16.50	4.08	13.50	60.38	13.50	46.88	—	4.08
8	—	—	—	(9.41)	(2.29)	16.50	4.80	14.21	46.88	11.72	35.16	—	4.80
9	—	—	—	(9.41)	(1.59)	16.50	5.50	14.91	35.16	8.79	26.37	(1.25)	4.25
10	—	—	—	(9.41)	(0.88)	16.50	6.21	15.62	26.37	6.59	19.78	(3.06)	3.15
11	—	—	—	(4.71)	(0.18)	16.50	11.61	16.32	19.78	4.94	14.84	(4.51)	7.10
12	—	—	—	—	—	16.50	16.50	16.50	14.84	3.71	11.13	(5.69)	10.81
13	—	—	—	—	—	16.50	16.50	16.50	11.13	2.78	8.35	(6.40)	10.10
14	—	—	—	—	—	16.50	16.50	16.50	8.35	2.09	6.26	(6.86)	9.64
15	—	—	—	—	—	16.50	16.50	16.50	6.26	1.57	4.69	(7.20)	9.30
16	—	—	—	—	—	16.50	16.50	16.50	4.69	1.17	3.52	(7.47)	9.03
17	5.00	—	—	—	—	16.50	21.50	21.50	3.52	3.52	—	(7.66)	13.84
18	—	—	—	—	—	—	—	—	—	—	—	(8.99)	(8.99)
	(95.00)	(3.00)	80.00	(80.00)	(31.32)	247.50	118.18	221.18	—	103.00	—	(59.09)	59.09

Table 18

Year	Capital Expenditure	Other non-operating exp. & income	Loan drawdowns	Capital repayments	Interest repayments	Net charter income	Net cash flow before tax	Profits for tax	Cum. capital allowances b/f	Capital allowances used	Capital allowances c/f	Tax at 50%	Net cash flow after tax
0	(5.00)	(1.00)	—	—	—	—	(6.00)	—	—	6.00	—	—	(6.00)
1	(40.00)	—	35.00	—	—	—	(5.00)	—	—	40.00	—	3.00	(2.00)
2	(55.00)	(2.00)	45.00	—	(4.32)	—	(16.32)	(4.32)	—	57.00	—	20.00	3.68
3	—	—	—	(9.41)	(5.82)	16.50	1.27	10.68	—	—	—	30.66[1]	31.93
4	—	—	—	(9.41)	(5.12)	16.50	1.97	11.38	—	—	—	(5.34)	(3.37)
5	—	—	—	(9.41)	(4.41)	16.50	2.68	12.09	—	—	—	(5.69)	(3-01)
6	—	—	—	(9.41)	(3.71)	16.50	3.38	12.79	—	—	—	(6.05)	(2.67)
7	—	—	—	(9.42)	(3.00)	16.50	4.08	13.50	—	—	—	(6.39)	(2.31)
8	—	—	—	(9.41)	(2.29)	16.50	4.80	14.21	—	—	—	(6.75)	(1.95)
9	—	—	—	(9.41)	(1.59)	16.50	5.50	14.91	—	—	—	(7.11)	(1.61)
10	—	—	—	(9.41)	(0.88)	16.50	6.21	15.62	—	—	—	(7.45)	(1.24)
11	—	—	—	(4.71)	(0.18)	16.50	11.61	16.32	—	—	—	(7.81)	3.80
12	—	—	—	—	—	16.50	16.50	16.50	—	—	—	(8.16)	8.34
13	—	—	—	—	—	16.50	16.50	16.50	—	—	—	(8.25)	8.25
14	—	—	—	—	—	16.50	16.50	16.50	—	—	—	(8.25)	8.25
15	—	—	—	—	—	16.50	16.50	16.50	—	—	—	(8.25)	8.25
16	—	—	—	—	—	16.50	16.50	16.50	—	—	—	(8.25)	8.25
17	5.00	—	—	—	—	16.50	21.50	21.50	—	—	—	(8.25)	13.25
18	—	—	—	—	—	—	—	—	—	—	—	(10.75)	(10.75)
	(95.00)	(3.00)	80.00	(80.00)	(31.32)	247.50	118.18	221.18	—	103.00	—	(59.09)	59.09

[1] Assumes allowances on interest paid during construction can also be used elsewhere.

Table 19

Year	Cash flow after tax	Present value value factor at 68.17%	Present value	Cumulative present value
0	(6.00)	1.0000	(6.000)	(6.000)
1	(2.00)	0.5946	(1.189)	(7.189)
2	3.68	0.3536	1.301	(5.888)
3	31.93	0.2103	6.715	0.827
4	(3.37)	0.1250	(0.421)	0.406
5	(3.01)	0.0743	(0.224)	0.182
6	(2.67)	0.0442	(0.118)	0.064
7	(2.31)	0.0263	(0.061)	0.003
8	(1.95)	0.0156	(0.031)	(0.028)
9	(1.61)	0.0093	(0.015)	(0.043)
10	(1.24)	0.0055	(0.007)	(0.050)
11	3.80	0.0033	0.012	(0.038)
12	8.34	0.0020	0.016	(0.022)
13	8.25	0.0012	0.010	(0.012)
14	8.25	0.0007	0.006	(0.006)
15	8.25	0.0004	0.003	(0.003)
16	8.25	0.0002	0.002	(0.001)
17	13.25	0.0002	0.002	0.001
18	(10.75)	0.0001	(0.001)	0
	59.09			

The examples considered have ignored the possibility that the capital allowances on the new ship could be used to reduce (or perhaps eliminate) the owners tax liability on other profits. In this case the changes in the tax payable on other income should be taken into account in the cash flows for the new project in order that the full effect of embarking on the project is known. Let us suppose therefore that the owner has large profits from other sources that enable him to make full use of the allowances as they become available, other factors remaining unchanged. The effect is shown in the following Table 18 which uses the same columns as Table 16 to facilitate comparison.

The IRR calculated on the above figures is 68.17 per cent! Examination of the cash flows shows a rather abnormal pattern however because the combination of the loan funding and the tax saving result in the project having cash in hand in excess of the project requirements in the early years. Applying the test given in Chapter 6 (page 42) it will be seen that this is a case where the use of the usual method implies assumptions which are not acceptable. See Table 19.

Faced with this situation the analyst has four alternatives open to him:—

1. Eliminate the cash surplus in year 3 as described below.
2. Ignore the loan funding in the calculation.
3. Follow the method set out in Appendix 1.
4. Ignore the DCF method altogether.

To follow the first proposal we use the method already described to eliminate a negative cash flow at the end of a project; the underlying principles being the same in each case. If we consider the practical situation where there are positive cash flows preceding negative cash flows and where money in hand invested produces a return below the IRR, it is reasonable to assume that the positive cash flows would

Table 20

	N.C.F. after tax	Modified cash flow
0	(6.00)	(6.00)
1	(2.00)	(2.00)
2	3.68	3.68
3	31.93	18.17
4	(3.37)	—
5	(3.01)	—
6	(2.67)	—
7	(2.31)	—
8	(1.95)	—
9	(1.61)	—
10	(1.24)	—
11	3.80	3.80
12	8.34	8.34
13	8.25	8.25
14	8.25	8.25
15	8.25	8.25
16	8.25	8.25
17	13.25	3.01
18	(10.75)	—

IRR = 49.29%

be invested to cover following negative cash flows (which must exist if the IRR is correct since there will be nil capital outstanding at the end of the project). Thus in the example (Table 18) the positive cash flow in years 2 and 3 would be invested to cover the following negative cash flows.

In a practical case it is not necessary to go to the trouble of Table 19; the situation in such a case can be quickly tested by taking cumulative PV's from year 0 and seeing whether they become positive before the cash flows become negative again, i.e. the first four lines of Table 19 are all that is necessary. (With some pocket calculators which calculate PV's based on assumed IRR's by cumulating the PV's from the last period of change of sign in the total will indicate that the IRR [assuming the final PV = 0] is fallacious).

Considering the practical case, having established special treatment is required, the initial cash flows must be modified by taking the cash surplus at the end of year 3 and using this to reduce or eliminate negative cash flows. This is illustrated in Table 20 where investment of the surplus in year 3 at five per cent pa has been assumed.

Thus the owner, otherwise liable for heavy tax, who can make maximum use of the most favourable system of capital allowances is far better off than the owner working in a no-tax regime. This is because the owner's investment is eliminated in year 3 by the tax saving which is in effect an interest free loan from the government! The IRR is however based on small investments for a short period and becomes very misleading. Small changes in income would produce violent changes in the IRR. For example a 10 per cent drop in the net income would reduce the IRR to 35 per cent. Thus the use of this type of analysis can only be justified when the income is guaranteed.

But what about the owner with no initial allowances but with plenty of taxable profits? His position is shown in Table 21 below; the IRR is 21.50 per cent compared to the "no-tax" position of 12.27 per cent (Table 14) an annual writing down allowance of 25 per cent being assumed.

Table 21

Year	Capital Expenditure	Other non-operating exp. & income	Loan drawdowns	Capital repayments	Interest repayments	Net charter income	Net cash flow before tax	Profits for tax	Cum. capital allowances b/f	Capital allowances used	Capital allowances c/f	Tax at 50%	Net cash flow after tax
0	(5.00)	(1.00)	—	—	—	—	(6.00)	—	—	1.50	4.50	—	(6.00)
1	(40.00)	—	35.00	—	—	—	(5.00)	—	4.50	11.13	33.38	0.75	(4.25)
2	(55.00)	(2.00)	45.00	—	(4.32)	—	(16.32)	(4.32)	33.38	22.59	67.78	5.57	(10.75)
3	—	—	—	(9.41)	(5.82)	16.50	1.27	10.68	67.78	16.95	50.84	13.46[1]	14.73
4	—	—	—	(9.41)	(5.12)	16.50	1.97	11.38	50.84	12.71	38.13	3.14	5.11
5	—	—	—	(9.41)	(4.41)	16.50	2.68	12.09	38.13	9.53	28.60	0.67	3.35
6	—	—	—	(9.41)	(3.71)	16.50	3.38	12.79	28.60	7.15	21.45	(1.28)	2.10
7	—	—	—	(9.42)	(3.00)	16.50	4.08	13.50	21.45	5.36	16.08	(2.82)	1.26
8	—	—	—	(9.41)	(2.29)	16.50	4.80	14.21	16.08	4.02	12.06	(4.07)	0.73
9	—	—	—	(9.41)	(1.59)	16.50	5.50	14.91	13.06	3.02	9.05	(5.10)	0.40
10	—	—	—	(9.41)	(0.88)	16.50	6.21	15.62	9.05	2.26	6.79	(5.95)	0.26
11	—	—	—	(4.71)	(0.18)	16.50	11.61	16.32	6.79	1.70	5.09	(6.68)	4.93
12	—	—	—	—	—	16.50	16.50	16.50	5.09	1.27	3.82	(7.31)	9.19
13	—	—	—	—	—	16.50	16.50	16.50	3.82	0.95	2.86	(7.62)	8.88
14	—	—	—	—	—	16.50	16.50	16.50	2.86	0.72	2.15	(7.78)	8.72
15	—	—	—	—	—	16.50	16.50	16.50	2.15	0.54	1.61	(7.89)	8.61
16	—	—	—	—	—	16.50	16.50	16.50	1.61	0.40	1.21	(7.98)	8.52
17	5.00	—	—	—	—	16.50	21.50	21.50	1.21	1.21	—	(8.05)	13.45
18	—	—	—	—	—	—	—	—	—	—	—	(10.15)	(10.15)
	(95.00)	(3.00)	80.00	(80.00)	(31.32)	247.50	118.18	221.18	—	103.01	—	(59.09)	59.09

(1) Assumes allowance on interest can also be used elsewhere.

Table 22

Year	Capital Expenditure	Other non-operating exp. & income	Loan drawdowns	Capital repayments	Interest repayments	Net charter income	Net cash flow before tax	Profits for tax	Cum. capital allowances b/f	Capital allowances used	Capital allowances c/f	Tax at 50%	Net cash flow after tax
0	(5.00)	(1.00)	—	—	—	—	(6.00)	—	—	—	6.00	—	(6.00)
1	(40.00)	—	35.00	—	—	—	(5.00)	—	(6.00)	—	46.00	—	(5.00)
2	(55.00)	(2.00)	45.00	—	(4.32)	—	(16.32)	(4.32)	46.00	—	103.00	—	(16.32)
3	—	—	—	(9.41)	(5.82)	16.50	1.27	10.68	103.00	25.75	77.25	—	1.27
4	—	—	—	(9.41)	(5.12)	16.50	1.97	11.38	77.25	19.31	57.94	9.70	11.67
5	—	—	—	(9.41)	(4.41)	16.50	2.68	12.09	57.94	14.48	43.46	3.96	6.64
6	—	—	—	(9.41)	(3.71)	16.50	3.38	12.79	43.45	10.86	32.60	1.20	4.58
7	—	—	—	(9.42)	(3.00)	16.50	4.08	13.50	32.60	8.15	24.45	(0.96)	3.12
8	—	—	—	(9.41)	(2.29)	16.50	4.80	14.21	24.45	6.11	18.34	(2.68)	2.12
9	—	—	—	(9.41)	(1.59)	16.50	5.50	14.91	18.34	4.58	13.75	(4.05)	1.45
10	—	—	—	(9.41)	(0.98)	16.50	6.21	15.62	13.76	3.44	10.32	(5.16)	1.05
11	—	—	—	(4.71)	(0.18)	16.50	11.61	16.32	10.32	2.58	7.74	(6.09)	5.52
12	—	—	—	—	—	16.50	16.50	16.50	7.74	1.93	5.81	(6.87)	9.63
13	—	—	—	—	—	16.50	16.50	16.50	5.81	1.45	4.36	(7.29)	9.21
14	—	—	—	—	—	16.50	16.50	16.50	4.36	1.09	3.27	(7.52)	8.98
15	—	—	—	—	—	16.50	16.50	16.50	3.27	0.82	2.45	(7.71)	8.79
16	—	—	—	—	—	16.50	16.50	16.50	2.45	0.61	1.84	(7.84)	8.66
17	5.00	—	—	—	—	16.50	21.50	21.50	1.84	1.84	—	(7.95)	13.55
18	—	—	—	—	—	—	—	—	—	—	—	(9.83)	(9.83)
	(95.00)	(3.00)	(80.00)	(80.00)	(31.32)	242.50	118.18	221.18	—	103.00	—	(59.09)	59.09

At this point it is worth considering the value of being able to utilise the capital allowances on the instalments as they are paid, rather than waiting until delivery of the vessel. The latter situation is shown in table 22, and shows an IRR of 15.59 per cent compared to the earlier figure of 21.50 per cent, showing the importance of the timing of early cash flows in a project. It is also worth noting that in all the com-

Table 23 No tax situation

Year	Capital expenditure	Other expenditure and income	Net charter income	Net cash flow (no tax)
0	(5.00)	(1.00)	—	(6.00)
1	(40.00)	—	—	(40.00)
2	(55.00)	(2.00)	—	(57.00)
3	—	—	16.50	16.50
4	—	—	16.50	16.50
5	—	—	16.50	16.50
6	—	—	16.50	16.50
7	—	—	16.50	16.50
8	—	—	16.50	16.50
9	—	—	16.50	16.50
10	—	—	16.50	16.50
11	—	—	16.50	16.50
12	—	—	16.50	16.50
13	—	—	16.50	16.50
14	—	—	16.50	16.50
15	—	—	16.50	16.50
16	—	—	16.50	16.50
17	5.00	—	16.50	21.50
18	—	—	—	—
	(95.00)	(3.00)	247.50	149.50

IRR = 12.61%

parisons made above the ammount of tax remains the same — it is only the timing of the payments that differs.

In making comparisons between different tax systems, the appropriate rate of tax must, of course, be used in each case. If the calculation in Table 21 was being which lowers the IRR to 19.48 per cent. At first sight this appears odd — one would expect a lowering of the tax rate to increase the IRR. However, lowering the tax rate reduces the amount of the tax recovered in the early years and this more than offsets the lower, but heavily discounted, reduced payments of tax in later years.

If we ignore the loan funding in the projects then the comparison between the three situations we have considered above (i.e. no tax, isolated project, and maximum use of allowances) would be as shown in the following tables, all other factors remaining the same (see page 44).

Table 24 Isolated project (most favourable regime)

Year	Capital expenditure	Other expenditure and income	Net charter income	Net cash flow before tax	Allowances in year	Cumulative allowances	Allowances used in year	Allowances carried forward	Tax at 50%	Net cash flow after tax
0	(5.00)	(1.00)	—	(6.00)	6.00	6.00	—	6.00	—	(6.00)
1	(40.00)	—	—	(40.00)	40.00	64.00	—	46.00	—	(40.00)
2	(55.00)	(2.00)	—	(57.00)	57.00	103.00	—	103.00	—	(57.00)
3	—	—	16.50	16.50	—	103.00	16.50	86.50	—	16.50
4	—	—	16.50	16.50	—	86.50	16.50	70.00	—	16.50
5	—	—	16.50	16.50	—	70.00	16.50	53.50	—	16.50
6	—	—	16.50	16.50	—	53.50	16.50	37.00	—	16.50
7	—	—	16.50	16.50	—	37.00	16.50	20.50	—	16.50
8	—	—	16.50	16.50	—	20.50	16.50	4.00	—	16.50
9	—	—	16.50	16.50	—	4.00	4.00	—	—	16.50
10	—	—	16.50	16.50	—	—	—	—	(6.25)	10.25
11	—	—	16.50	16.50	—	—	—	—	(8.25)	8.25
12	—	—	16.50	16.50	—	—	—	—	(8.25)	8.25
13	—	—	16.50	16.50	—	—	—	—	(8.25)	8.25
14	—	—	16.50	16.50	—	—	—	—	(8.25)	8.25
15	—	—	16.50	16.50	—	—	—	—	(8.25)	8.25
16	—	—	16.50	16.50	—	—	—	—	(8.25)	8.25
17	5.00	—	16.50	21.50	—	—	—	—	(8.25)	13.25
18	—	—	—	—	—	—	—	—	(10.75)	(10.75)
	(95.00)	(3.00)	247.50	149.50					(74.75)	74.75

IRR = 8.98%   Note: in both this and the following table the IRR has been calculated after eliminating the negative cash flow in year 18 as already described.

Table 25
Maximum use of allowances (most favourable regime)

Year	Capital expenditure	Other expenditure and income	Net charter income	Net cash flow before tax	Allowances in year	Cumulative allowances	Allowances used in year	Allowances carried forward	Tax at 50%	Net cash flow after tax
0	(5.00)	(1.00)	—	(6.00)	6.00	6.00	6.00	—	—	(6.00)
1	(40.00)	—	—	(40.00)	40.00	40.00	40.00	—	3.00	(37.00)
2	(55.00)	(2.00)	—	(57.00)	57.00	57.00	57.00	—	20.00	(37.00)
3	—	—	16.50	16.50	—	—	—	—	28.50	45.00
4	—	—	16.50	16.50	—	—	—	—	(8.25)	8.25
5	—	—	16.50	16.50	—	—	—	—	(8.25)	8.25
6	—	—	16.50	16.50	—	—	—	—	(8.25)	8.25
7	—	—	16.50	16.50	—	—	—	—	(8.25)	8.25
8	—	—	16.50	16.50	—	—	—	—	(8.25)	8.25
9	—	—	16.50	16.50	—	—	—	—	(8.25)	8.25
10	—	—	16.50	16.50	—	—	—	—	(8.25)	8.25
11	—	—	16.50	16.50	—	—	—	—	(8.25)	8.25
12	—	—	16.50	16.50	—	—	—	—	(8.25)	8.25
13	—	—	16.50	16.50	—	—	—	—	(8.25)	8.25
14	—	—	16.50	16.50	—	—	—	—	(8.25)	8.25
15	—	—	16.50	16.50	—	—	—	—	(8.25)	8.25
16	—	—	16.50	16.50	—	—	—	—	(8.25)	8.25
17	5.00	—	16.50	21.50	—	—	—	—	(8.25)	13.25
18	—	—	—	—	—	—	—	—	(10.75)	(10.75)
	(95.00)	(3.00)	247.50	149.50	103.00				(74.75)	74.75

IRR 12.58% Thus in this instance, if maximum use can be made of the allowances by setting them against tax liabilities the IRR is almost exactly the same as in the 'no-tax' situation.

Chapter 9

Leasing

We have seen in the previous chapter how an owner who can use up the capital allowances on a new ship against other profits in the first year or so of a vessel's life has a considerable financial advantage over the owner with no other profits, both in terms of the capital required (in addition to the shipbuilding loan) and in terms of the rates of return on the capital employed (although in making the latter comparison remember that if the traditional DCF analysis is employed the same risks are being taken by differing amount of capital).

However an owner who cannot make rapid use of the capital allowances does have an alternative — he can arrange for another company, who can make immediate use of the allowances to own the ship and to lease it to him on a bare-boat charter whereby the actual owner or 'lessor' is involved in a purely financial operation, the charter terms ensuring that all the usual shipowning risks and liabilities are taken by the shipping company (the 'lessee'). The clauses which ensure this are sometimes called 'hell and high water clauses'. One important clause (or clauses) requires the owner to insure the vessel for at least certain minimum values which are stated and arranges for insurance monies recovered to be paid to the lessor.

The hire payable by the lessee will reflect the tax saving by the lessor, who is usually a bank or group of banks. Thus a leasing deal has two major advantages to the shipping company — firstly no capital outlay is required — the first payment in respect of the ship, apart from legal fees, being the first hire payment which would normally be due on the delivery of the ship — and secondly the rate of hire will reflect immediate use of the tax allowances.

The lessor as owner will qualify for the usual shipbuilding credit and will make all the pre-delivery payments including loan interest, commitment fees, etc. which will be 'rolled-up' into a total cost at delivery upon which the rate of hire to the lessee will be calculated.

The shipping company's cash flow will probably be further improved in the initial years of a lease because the hire payments are usually in equal amounts throughout the period of the charter whereas the loan payments (capital plus interest) by an owner are greatest in the early years because of the greater interest payments. Indeed there is no reason why the hire payments should not be 'stepped' with lower payments in the early years if lessor and lessee agree. Even without 'stepped' hire it is likely that the earnings reported in the Profit and Loss account in the early years will be greater if the vessel is leased because the hire for say a fifteen year lease will probably be less than depreciation plus the interest payments on a shipbuilding loan.

Other advantages of leasing are that it is not restricted by limitations on a company's borrowing power; it avoids possible dilution of capital that would arise if funds for ownership were raised by the sale of shares. In many countries lease transactions are not required to be noted on the Balance Sheet, and leasing is an 'off balance sheet' form of financing.

The disadvantages of a lease, as compared to direct ownership, are relatively minor. Although the lease may be for ten or fifteen years it is usual to have a clause in the lease agreement (a charter) which enables the lessee to have the vessel sold at any time after an initial period has elapsed (e.g. four or five years) and the difference, or the bulk of it, between the proceeds of the sale and the book value tabulated in the lease agreement, passed to him as rebate of hire. Thus the shipping company is not at any great disadvantage through leasing rather than ownership, in that if circumstances arise such that the lessee wishes to sell the ship he can do so. It should be noted that most tax regimes do not allow ownership of the vessel to pass to the lessee at the end of the lease, or indeed at any time without the lessor losing the capital allowances. It is therefore usual for there to be provision to continue the lease after the initial period of perhaps ten or fifteen years on a similar basis but with a 'peppercorn' rate of hire which might be one tenth of the rate during the initial period.

One important point about a lease is that although the bank or organisation actually owning the vessel has it as security, the value that it places on it will only be similar to the valuation it would make for mortgage purposes. The bank will therefore have to be satisfied that the owner will be able to make the hire payments when they fall due, which rather restricts the use of leasing to companies of good financial standing who are thereby less likely to be in a position to benefit fully from a lease such as has been described. The lessor may require a lien on all charter hires earned by the vessel.

It should be mentioned that where such 'tax' leases are concerned there are in some cases restrictions on chartering leased vessels to overseas companies in order to prevent the benefit of the tax allowances being enjoyed by an overseas company under conditions where tax is unlikely to be paid i.e. the export of tax allowances is not allowed. Regulations in the UK for example prevent such a practice.

Not all leases are tax orientated. There are occasions when a lease, which can be a very flexible financial instrument, is attractive to a company quite apart from tax considerations. Very often such leases are based on vessels owned by a 'flag of convenience'; the construction may be arranged in a country offering particularly attractive credit terms etc. — the opportunities for advantageous arrangements are sometimes considerable using such 'offshore' or 'cross-boundary' leases.

Leasing can be applied to second hand vessels but of course a tax lease is of no advantage in countries that do not allow accelerated use of capital allowances on such vessels.

In the case of new vessels there is usually a supervision agreement between lessor and lessee whereby the latter supervises the construction of the vessel in exactly the same way as he would if the vessel was in his direct ownership.

Table 26

Year	Capital Expenditure	Other non-operating expenditure	Loan drawdowns	Capital repayments	Interest repayments	Net charter income	Net cash flow before tax	Profits for tax	Capital allowances b/f	Capital allowances used (1)	Capital allowances c/f	Tax at 50%	Net cash flow after tax	NCF modified for analysis
0	(5.00)	(1.00)	—	—	—	—	(6.00)	—	—	6.00	—	—	(6.00)	(6.00)
1	(40.00)	—	35.00	—	—	—	(5.00)	—	—	40.00	—	2.55	(2.45)	(2.45)
2	(55.00)	(2.00)	45.00	—	(4.32)	—	(16.32)	(4.32)	—	57.00	—	17.00	0.68	0.68
3	—	—	—	(9.41)	(5.82)	11.94	(3.29)	6.12	—	—	—	26.39(2)	23.10	10.47
4	—	—	—	(9.41)	(5.12)	11.94	(2.59)	(6.82)	—	—	—	(3.06)	(5.65)	—
5	—	—	—	(9.41)	(4.41)	11.94	(1.88)	7.53	—	—	—	(3.41)	(5.29)	—
6	—	—	—	(9.41)	(3.71)	11.94	(1.18)	8.23	—	—	—	(3.76)	(4.94)	(2.10)
7	—	—	—	(9.42)	(3.00)	11.94	(0.48)	8.94	—	—	—	(4.12)	(4.60)	(4.60)
8	—	—	—	(9.41)	(2.29)	11.94	0.24	9.65	—	—	—	(4.47)	(4.23)	(4.23)
9	—	—	—	(9.41)	(1.59)	11.94	0.94	10.35	—	—	—	(4.82)	(3.88)	(3.88)
10	—	—	—	(9.41)	(0.88)	11.94	1.65	11.06	—	—	—	(5.18)	(3.53)	(3.53)
11	—	—	—	(4.71)	(0.18)	11.94	7.05	11.76	—	—	—	(5.53)	1.52	1.52
12	—	—	—	—	—	11.94	11.94	11.94	—	—	—	(5.88)	6.06	6.06
13	—	—	—	—	—	11.94	11.94	11.94	—	—	—	(5.97)	5.97	5.97
14	—	—	—	—	—	11.94	11.94	11.94	—	—	—	(5.97)	5.97	5.97
15	—	—	—	—	—	11.94	11.94	11.94	—	—	—	(5.97)	5.97	5.97
16	—	—	—	—	—	11.94	11.94	11.94	—	—	—	(5.97)	5.97	5.97
17	5.00	—	—	—	—	11.94	16.94	16.94	—	—	—	(5.97)	10.97	2.90
18	—	—	—	—	—	—	—	—	—	—	—	(8.47)	(8.47)	—
	(95.00)	(3.00)	80.00	(80.00)	(31.32)	179.10	49.78	152.78	—	103.00	—	(32.61)	17.17	—

Notes: (1) Capital allowances taken at 85% of value when calculating tax credit. (2) Assumes that interest paid during construction can be offset against other taxable profits.

We can now examine the actual operation of a lease by looking at it first from the lessor's point of view. Let us suppose that he requires 11 per cent pa IRR after tax on his own capital and that he retains 15 per cent of the tax saved as his profit from the deal. He can then calculate the hire — most leasing companies have computer programmes available which will do this without effort, otherwise it is necessary to perform the type of calculation shown in Table 26 for a range of hires (i.e. net charter income). Note that because all the shipowning risks are taken by the lessee and assuming he is of sound financial status the income is assured and therefore this form of analysis can be used with some confidence.

Table 26 (page 73) shows the principles for calculating a 15 year lease hire — it is of course a similar table to table 18 except that only 85 per cent of the tax credits have been taken.

Again there is the problem of the surplus in year 3, which we can deal with in a similar manner as before in Chapter 8. However then we assumed that the initial investment was not repaid from the surplus which was used to cover the following negative cash flows. In the leasing case, however, (especially if the lessor is a bank working on borrowed money) it may be more realistic to assume that this initial investment would be repaid and this assumption can be easily built into the calculation if we are seeking a particular IRR, since we can calculate what sum would be required in year 3 to repay this investment and then use the balance left to deal with following negative cash flows. This treatment has been applied in the example and the result shown as the last column in table 26.

The IRR found from the last column is just over 11.0 per cent pa. This IRR would be improved if there was the possibility of carrying back any of the allowances and obtaining the tax credits earlier. A residual value of 5.00 has been assumed — many leasing organisations would ignore the residual value but this has little effect on the hire rate for a given IRR because of the discounting.

The calculation in Table 26 has assumed that the allowances become available and can be used, as soon as each building instalment is paid, and furthermore that such allowances can be used in full, i.e. there is a 100 per cent initial allowance. If there are smaller initial allowances (or no initial allowances) the rate of hire required will increase. The changes in the UK system of capital allowances for ships which will be fully implemented in 1986 considerably increase the hire required from the lessee compared to the previous "free depreciation" era with 100 per cent first year allowances.

Examination of the net cash flows after tax (last but one column table 26) shows one of the problems faced by lessors in arranging a tax lease. Suppose we calculate the cumulative net cash flow after tax both as a simple addition and then with interest charged as shown in Table 27.

The last column shows that while the lessor requires relatively small funds to pay for the vessel initially, and will have cash in hand from years 3 to 6, thereafter he is required to inject funds each year until the shipbuilding credit has been repaid — furthermore the total capital outstanding at this point is substantially greater than the initial funds required. Thus if the lessor offers a fixed rate of hire throughout the period of the lease, he has to estimate the cost of the funds which will be required at periods of over ten years ahead in the case of a 15 year lease. Such estimates are likely to be conservative (i.e. to the lessor's advantage) and this problem has led to some leases having a variable rate based on the borrowing rates e.g. on LIBOR (London Inter Bank Offer Rate) — an independant measure of the rates at which money is lent and borrowed for different periods at a given time on a given day.

Table 27

Year	Net cash flow (NCF)	Cumulative NCF (no interest)	Cumulative NCF with interest (1)
0	(6.00)	(6.00)	(6.00)
1	(2.45)	(8.45)	(9.12)
2	0.68	(7.77)	(9.45)
3	23.10	15.33	12.60
4	(5.65)	9.68	7.58
5	(5.29)	4.39	2.67
6	(4.94)	(0.55)	(2.14)
7	(4.60)	(5.15)	(6.97)
8	(4.23)	(9.38)	(11.98)
9	(3.88)	(13.26)	(17.19)
10	(3.53)	(16.79)	(22.63)
11	1.52	(15.27)	(23.62)
12	6.06	(9.21)	(20.19)
13	5.97	(3.24)	(16.46)
14	5.97	2.73	(12.32)
15	5.97	8.70	(7.72)
16	5.97	14.67	(2.61)
17	10.97	25.64	8.07
18	(8.47)	17.17	0

Note: (1) With interest at 11.1% on sums outstanding and 5% on sums in hand.

Another important point is that in the calculations we have made e.g. Table 26 we have assumed that the rate of tax is unchanged throughout the period of the lease. If the rate of tax changes after the allowances have been used then the lessor will pay more or less tax than he calculated. For example some leasing organisations in the UK having assumed a constant rate of tax, suffered heavy losses when the rate of tax increased from 40 per cent to 50 per cent consequent upon a change in the tax systems. As a result it is now usual for tax based leases to incorporate a clause to adjust the rate of hire if the rate of tax changes — this can of course work either way.

In considering or comparing leases attention must also be paid to the clauses enabling the vessel to be sold if the lessee wishes. Although the lessor passes all 'shipowning' risks to the lessee, some lessors claim a part of the profits from the sale of the vessel, which might be legitimately considered to arise from the 'shipowning risks'. If this is the case it would be reasonable to expect it to be reflected in lower requirements regarding guarantees from the lessee.

When considering the appraisal of a leasing transaction as an investment it is obvious that the usual DCF methods cannot be applied since there is no initial investment. However the shipowner will usually be concerned with choosing between owning a vessel or alternatively leasing it, and in this situation there is no difficulty in principle in analysis. The purchase of the vessel can be regarded as effecting the saving of the lease payments i.e. the investment in the purchase of the vessel will increase the project's subsequent profits by the amount of the lease payments. The IRR found in this way should satisfy the same criteria as the company applies to its investments i.e. be at least a certain rate. This method is in fact the same as that used to compare two mutually exclusive investments as described in Chapter 6 (page 43).

In comparing a lease with ownership it is important to compare like for like. For example if the lease is for ten years, after which the lessee is left with nothing, then when comparing this with the cost of ownership the value of the vessel after ten years in the latter case must be taken into account.

Chapter 10
Some Relevant Matters

1. Inflation

In our considerations so far, particularly in respect of investment appraisal, we have assumed that the value of money remains unchanged with the passage of time — in other words a Pound Sterling will buy the same quantity of goods and services in ten years time as it would today. This may have been the case many years ago but it has certainly not been the case in recent years in any major currency.

We have therefore to consider the impact of the changing value of the unit of currency on financing arrangements and their appraisal. In some areas this can be a complex matter and the accountancy profession is far from agreement on the way in which inflation should be handled in the preparation of annual accounts, although a trial 'Standard' has been produced in order that companies can produce accounts which do allow for the effect of inflation in certain ways. A discussion of inflation accounting is outwith the scope of this book, but it can be said the most important differences between the accounts of a shipowner prepared on a historical basis and those accounts prepared on an inflation accounting basis is that the latter will provide much larger sums for depreciation (especially of course on older ships) and thus show lower profits while the following real value of loans will tend to increase profits. We must however consider the effect of inflation on our methods of fianance and on our methods of appraisal.

In discussions regarding inflation the expressions "real terms" and "real values" are used to distinguish between values expressed in unchanging units, as opposed to 'money values' where the value of the monetary unit is changing. Thus 'real terms' may make use of say "1980 £'s" as a measure of value, while money values are expressed in 'money of the day'! To avoid confusion it often helps to write for example 1980 £s as £80 etc.

To illustrate the use of these terms, suppose that in 1970 a company invested £1000 at 15 per cent pa for five years — it would then have received £2011 in 1975 (original capital plus compounded interest). Thus if there had been no inflation between 1970 and 1975 the investment would have doubled in value. In fact however there was severe inflation and the value of the Pound in 1975 was only 54 per cent of its value in 1970. Thus we could say that in real terms the value of the investment had increased to

$$£_{70}(2011x.54) = £_{70}1086$$

an increase of 8.6 per cent on the original. Therefore the value increased 8.6 per cent in real terms or 101 per cent in money terms.

From the point of view of the lender in this example, inflation almost wiped out his gain from the interest paid — indeed there have been periods in a number of countries where the rate of inflation has exceeded the going rates of interest — but just as the lender has suffered so will the borrower have gained if his income is fixed in real rather than money terms. If he had calculated in 1970 that in 1975 he would have income of £2011 without allowing for the effects of inflation, then in fact he would have not £2011 available but

$$£_{75}\frac{(2011)}{.54} = £_{75}3,724$$

Having repaid the lender his £2011 the borrower would be left with a surplus of £1,713.

Thus there are considerable gains to be made by borrowing during periods of inflation in cases where income is determined in real terms and where the rate of interest has not risen to take full account of the rate of inflation. Inflation thus makes cheap shipbuilding loans look even more attractive. When considering other sources of funds the question has to be asked whether or not the interest/dividend payable and any other payments will be expected to be maintained in real terms or money terms — for example shareholders will expect their dividends (and thereby the value of their shares) to be maintained in real terms.

When we come to take account of inflation in investment appraisal, it is very much a case of considering each item of income or expenditure to determine whether it will be fixed in real terms or money terms. It is then necessary to convert all income and expenditure into either real terms or money terms before doing the appraisal calculations.

To illustrate this consider Table 16 (page 60). Column 3 shows the capital expenditure on the vessel during the building period and the amounts payable will have been fixed in the building contract in money terms. (There was a period in the early 1970's when shipbuilders in a strong seller's market introduced inflation linked prices but there seems little prospect of such times returning for many years!) Column 3 covers owner's supply, supervision etc. and the owner will know on which basis these estimates have been made. The sale of the vessel for scrap in year 17 in this column is expressed in real terms. Column 4, the loan drawdowns will obviously have been agreed in money terms, as will Columns 5 and 6, the capital repayments and the interest payments.

Column 7, the net charter income, may provide some difficulty. When there is a long term charter the terms of the charter will determine whether or not there is any adjustment for inflation in the rate of and, if so, whether this covers the whole hire or only a part, e.g. the direct operating costs. In this appraisal we are concerned with the net income after operating costs have been paid — if there is no provision in the charter for escalation of hire then, as inflation increases the operating costs in money terms, the amount left, i.e. the net income, will fall in money terms (and even more in real terms). This could be expressed another way by saying that if the hire is fixed in money terms it will fall in real terms and because the operating costs will be constant (we assume) in real terms the net income must fall in real terms.

On the other hand if the net income used in the appraisal is based on market rates from time to time i.e. there is no long term charter, then the position is less clear since the market rate will be determined by many factors. However we are concerned here only with inflation and must therefore ignore all other factors. The market rates will be influenced by operating and laying up costs in periods of low rates and by operating and new building costs in periods of higher rates. We can assume that these factors are all constant in real terms and therefore over a long period market rates will tend to reflect any changes in money values.

There is also the consideration that the income (and possibly other factors) are in a different currency. In Chapter 7 we considered how this problem could be dealt with but we ignored inflation. If we are due to say receive a sum in a foreign currency some years ahead how should we handle it in our investment appraisal? We may have the problem of both determining the rate of exchange as we have already discussed (Chapter 7) and considering the effects of inflation — the rates of inflation may be different in the two countries. However, it can be argued that over a period of time differences between the rates of inflation in two countries will be reflected in changes in the rate of exchange which will result from this difference. We can therefore convert the foreign currency as described earlier and then treat it for inflation ignoring its foreign origin.

Column 8, the cash flow before tax will have to be recalculated in either real terms or money terms since any mixture is meaningless. The next four columns dealing with tax allowances are based strictly on money terms, no government so far having granted capital allowances linked to an index of inflation! This further emphasises the advantage of using the allowances quickly before their value falls. The tax payable and the net cash flow after tax will then have to be calculated in the terms that have been chosen for the calculation.

The calculation in Table 16 can therefore be performed in real terms or money terms and this has been done assuming a rate of inflation of 5 per cent pa. Table 29 shows the calculation in money terms. Table 30 the same calculation in real terms. A comparison of the tables shows how in this case it is clearly easier to work in money terms — note that the totals of the columns in Table 29 are really meaningless since they are a summation of quantities in different units whereas the summation of the columns in Table 30 are meaningful since the individual quantities have all been measured in the same unit (in this case the value of the monetary unit in year 0).

The difference between the IRR derived from Table 29 and that from Table 30 is due simply to inflation — in money terms 100 in year 0 would become 125.38 in year 1 with interest at 25.38 per cent. But in year 1 we need 105 units of currency to have the same value as 100 in year 0, so our 125.38 in year 1 has the same value as $\frac{125.37}{105} = 119.40$ in year 0, i.e. in real terms our 100 units have increased by 19.40% (Table 30).

As with all analysis work, care must be taken to ensure that the basis for the calculations is clearly set out and that all comparisons are made on this same basis.

Table 29

Year	Real value of unit of currency	Capital Expenditure	Other non-operation exp. & income	Loan drawndowns	Capital repayments	Interest repayments	Net charter income	Net cash flow before tax	Profits for tax	Cum. capital allowances b/f	Capital allowances used	Capital allowances c/f	Tax at 50%	Net cash flow after tax
0	1.00	(5.00)	(1.00)	—	—	—	—	(6.00)	—	—	—	6.00	—	(6.00)
1	0.952	(40.00)	—	35.00	—	—	—	(5.00)	—	6.00	—	46.00	—	(5.00)
2	0.907	(55.00)	(2.00)	45.00	—	(4.32)	—	(16.32)	(4.32)	46.00	—	103.00	—	(16.32)
3	0.864	—	—	—	(9.41)	(5.82)	19.10	3.87	13.28	103.00	18.96	94.04	—	3.87
4	0.823	—	—	—	(9.41)	(5.12)	20.05	5.52	14.93	94.04	14.93	79.11	—	3.52
5	0.784	—	—	—	(9.41)	(4.41)	21.05	7.23	16.64	79.11	16.64	62.47	—	7.23
6	0.746	—	—	—	(9.41)	(3.71)	22.12	9.00	18.41	62.47	18.41	44.06	—	9.00
7	0.711	—	—	—	(9.42)	(3.00)	23.21	10.79	20.21	44.06	20.21	23.85	—	10.79
8	0.677	—	—	—	(9.41)	(2.29)	24.37	12.67	22.08	23.85	22.08	1.77	—	12.67
9	0.645	—	—	—	(9.41)	(1.59)	25.58	14.58	23.99	1.77	1.77	—	—	14.58
10	0.614	—	—	—	(9.41)	(0.88)	26.87	16.58	25.99	—	—	—	(11.11)	5.47
11	0.585	—	—	—	(4.71)	(0.18)	28.21	23.32	28.03	—	—	—	(13.00)	10.32
12	0.537	—	—	—	—	—	29.62	29.62	29.62	—	—	—	(14.01)	15.61
13	0.530	—	—	—	—	—	31.13	31.13	31.13	—	—	—	(14.81)	16.32
14	0.505	—	—	—	—	—	32.67	32.67	32.67	—	—	—	(15.57)	17.10
15	0.481	—	—	—	—	—	34.30	34.30	34.30	—	—	—	(16.33)	17.97
16	0.458	—	—	—	—	—	36.03	36.03	36.03	—	—	—	(17.15)	—
17	0.436	11.46	—	—	—	—	37.84	49.30	49.30	—	—	—	(18.05)	31.39
18	0.416	—	—	—	—	—	—	—	—	—	—	—	(24.65)	(24.65)
		(88.54)	(3.00)	80.00	(80.00)	(31.32)	412.42	289.56					114.78	144.78

Assumptions: 5% pa rate of inflation. Charter income assumed to be constant in real terms. IRR = 25.38%

Table 30

Year	Real value of unit of currency	Capital Expenditure	Other non-operation exp. & income	Loan drawndowns	Capital repayments	Interest repayments	Net charter income	Net cash flow before tax	Profits for tax	Cum. capital allowances b/f	Capital allowances used	Capital allowances c/f	Tax at 50%	Net cash flow after tax
0	1.000	(5.00)	(1.00)	—	—	—	—	(6.00)	—	—	—	6.00	—	(6.00)
1	0.952	(38.10)	—	33.32	—	—	—	(4.78)	—	5.71	—	43.81	—	(4.76)
2	0.907	(49.89)	(1.81)	40.82	—	(3.92)	—	(14.80)	(3.92)	41.72	—	97.34	—	(14.30)
3	0.864	—	—	—	(8.13)	(5.03)	16.50	3.34	11.47	88.97	7.74	81.23	—	3.34
4	0.823	—	—	—	(7.74)	(4.21)	16.50	4.55	12.29	77.36	12.29	65.08	—	4.55
5	0.784	—	—	—	(7.38)	(3.46)	16.50	5.66	13.04	61.97	13.04	48.94	—	5.66
6	0.746	—	—	—	(7.02)	(2.77)	16.50	6.71	13.73	46.60	13.73	32.88	—	6.71
7	0.711	—	—	—	(6.70)	(2.13)	16.50	7.67	14.37	31.31	14.37	16.94	—	7.67
8	0.677	—	—	—	(6.37)	(1.55)	16.50	8.58	14.95	16.13	14.95	1.18	—	8.58
9	0.645	—	—	—	(6.07)	(1.03)	16.50	9.40	15.47	1.12	1.12	—	—	9.40
10	0.614	—	—	—	(5.78)	(0.54)	16.50	10.18	15.96	—	—	—	(6.83)	3.35
11	0.585	—	—	—	(2.76)	(0.11)	16.50	13.63	16.39	—	—	—	(7.60)	6.03
12	0.557	—	—	—	—	—	16.50	16.50	16.50	—	—	—	(7.81)	8.69
13	0.530	—	—	—	—	—	16.50	16.50	16.50	—	—	—	(7.86)	8.64
14	0.505	—	—	—	—	—	16.50	16.50	16.50	—	—	—	(7.86)	8.64
15	0.481	—	—	—	—	—	16.50	16.50	16.50	—	—	—	(7.86)	8.64
16	0.458	—	—	—	—	—	16.50	16.50	16.50	—	—	—	(7.86)	8.64
17	0.436	5.00	—	—	—	—	16.50	21.50	16.50	—	—	—	(7.86)	13.64
18	0.416	—	—	—	—	—	—	—	—	—	—	—	(10.24)	(10.24)
		(87.99)	(2.81)	74.14	(57.95)	(24.75)	247.50	148.18					(71.78)	76.38

IRR = 19.40%

2. Working with others

Occasions sometimes arise when the possibility of sharing the ownership of a vessel appears attractive and the question arises as to determining the best manner by which this can be effected.

Suppose two companies wish for commercial reasons to share the ownership of a vessel the principle options are:—

a) a new company with a 50/50 shareholding to own the vessel.
b) a partnership with 50/50 interests.
c) direct ownership by each party of a one half share of the vessel (or whatever ratio is required if not 50/50).

Particular circumstances of the case may make one or more of these unacceptable e.g. one of the potential owners does not want a share in a ship to appear on his balance sheet, but otherwise the choice will usually be determined by firstly the laws relating to the legal liabilities of the individual owners in the different forms and secondly by taxation considerations. The choice will then depend upon the situation of the potential owners regarding their individual tax profits and losses relative to the capital allowances and anticipated profits from the new vessel, which will depend upon the relevant tax laws; here we shall give only a broad guide to the UK position in order to illustrate the considerations.

The first alternative, a new company, would mean that the ship was owned by a separate legal entity; the new company would itself be entitled to the capital allowances to offset against the trading profits and the profits of trading would be distributed to the owners by the payment of dividends. Although at first sight this might appear to rule out any use of the capital allowances on the vessel by the owners of the company there are special provisions — "Consortia relief" — which would enable a tax loss to be created in the new company by using the available allowances (or a part thereof) and such a loss can then be taken up by the separate owners. Such an arrangement can be applied provided that the new company is by UK registered companies each of which has at least a 5% stake. It is important to note however that losses cannot be passed the other way — if the owning companies are in or get into a loss making position such losses cannot be used to reduce the tax payable on any of the profits in the new company.

The fact that distribution to the owners would be by way of dividend in this first case does not present problems of Advance Corporation Tax (ACT) and tax credits because the new company would be able to make an election to pay dividends without givig rise to a liability for ACT and without tax credits.

The second alternative, a partnership, again means that the vessel will not be owned directly by the owners, but through the partnership which will utilise the capital allowances before the profit and loss is divided and passed to the owners in proportion to their share in the partnership. An important difference between this and the first alternative is that the profits or losses are regarded as a separate trade of the owners and if an owner has a trading loss this can be used to reduce the tax liability on his share of the profits of the new venture; on the other hand the use of the capital allowances to create a loss is determined by the partnership and not the individuals.

Partnerships in the UK have unlimited liability, and each partner is responsible for the liabilities of the whole partnership (if the other partners fail) except in the case of limited partners who take no part in the running of the partnership. Partnerships are therefore unpopular in the UK, but in some countries, the USA for example, a partnership can have limited liability.

The third alternative means the owners taking shares in the ship directly. In the UK as was mentioned in Chapter 1 the ownership of every merchant ship is divided in 64 and the direct ownership of a vessel by two or more parties is thereby facilitated. Each owner takes his direct share of all profits and liabilities from the venture and is entitled to his share of the capital allowances which because they go direct to the shareholders, can be used by each shareholder separately, as suits him, without consideration for the others.

This third alternative is thus the most flexible in so far as the taxation aspects are concerned. There should be no difficulty in securing cheap credit for a new vessel in any of the three cases — in the first it is probable that guarantees would be sought from the shareholders of the new company; in the third the documentation will be lengthy if there are several owners! In the case of the second alternative each partner would be responsible for the whole debts of the partnership i.e. the repayment of the whole loan which might be a consideration vis-a-vis any borrowing limits. In the first instance guarantees would undoubtedly be sought from the owners of the new company and these could well be 'joint and several' guarantees which mean in effect that each guarantor guarantees the whole loan.

It is worth mentioning here that in the UK up to five names can be registered as owners of each share, making it possible to register a ship with 320 owners! Similar types of situation exist in other countries and attempts have been made to set up groups of people with high incomes as shipowners in this way, the attraction to them being the use of capital allowances from their share of the ship to reduce or eliminate their high income tax. A merchant bank in the UK tried to organise such a group to own and lease ships, but legislation stopped the possibility, which would have been very attractive to taxpayers paying tax at the higher rates.

3. Improving the Contract

Before a shipbuilding contract is signed both owner and builder will endeavour to ensure that the contract is as attractive as possible to themselves. The contract must be attractive to both parties if it is to be signed; some cooperation can often produce a mutual benefit and therefore in this section attention is drawn to some considerations affecting one or both parties.

Where subsidies are available obviously maximum use must be made of them and expenditure not qualifying for subsidy kept to a minimum. Checks should also be made to ensure that it is not possible to increase the subsidy by additional expenditure that can be easily recouped at a later date — for instance if containers are included in the contract for a "container friendly" ship will they attract a shipbuilding subsidy which will be retained by the builder even if the containers are sold in new condition by the owner the day after delivery?

Similarly where cheap credit is available every effort should be made to ensure that as far as possible all expenditure qualifies for credit as has been described in Chapter 3 — for example the usual 'Owners supply' items to be purchased by the yard and included in the contract; a PC sum to cover extras, etc.

The builder will probably have arranged the instalments payable during construction to cover his outgoings. Any delay in the timing of instalments obviously helps the owner by reducing pre-delivery interest — while such delay will also show beneficially in the owner's financial analysis. The builder should obviously press his suppliers for credit terms which will enable him to present a more attractive instalment pattern to the owner. However it is not in the owners interest to risk the builder getting into financial difficulties during the construction of the vessel as a result of delaying instalments — there could be little worse for an owner than to have a half built, half paid for vessel, in a shipyard in liquidation — and this is not therefore a matter to press the builder over unless he has large financial resources or has a government guarantee. It should be remembered in this connection that state own shipyards do not necessarily have such a guarantee.

How much can be gained by delaying instalments? In Table 3 (p.19) a typical pattern of instalments is shown with 15 months between commencement of fabrication and delivery. Suppose this was reduced to 10 months — Table 3 would then look like this:-

Table 30

Event	Time from signing	Owners capital payments	Loan drawdown	Drawdowns interest on loan
Signing of contract	0 months	5	—	—
1,000 tons steel delivered	11 months	5	—	—
Commencement of steel fab.	14 months	—	15	—
25% steel erected	16 months	—	20	—
75% steel erected	19 months	3	17	—
—	20 months	—	—	1.17
Steel erection completed	21 months	3	12	—
Launch	22 months	1	4	—
Delivery	24 months	3	12	2.23
			80	3.40

Thus shortening the building time of the vessel (but leaving the overall time from contract to delivery unchanged) has reduced the interest payable on loan drawdowns by 0.92 — in addition there will be a benefit to the owner in respect of his own contributions. We can calculate the benefit to the owner by comparing the NPV of his payments in the two cases. The original schedule gives NPV's of 21.047 (15 per cent) and 20.184 (20 per cent) — the revised schedule gives 19.808 (15 per cent) and 18.870 (20 per cent). Thus the shortening of the building period (without changing the overall time from signing to delivery) reduces the NPV's by about 1.20. This would improve the IRR in the case considered by about 0.6 per cent from 17.3 per cent to 17.9 per cent. (Note that these NPV's have been calculated on a monthly basis with 15 per cent pa = 1.1715 per cent pm and 20 per cent pa = 1.5309 per cent pm.

Suppose that the overall delivery time was reduced by one third (rather than just the construction time). What effect will this have on cash flow and financial analysis? Let us assume that in Table 3 the steel fabrication starts 6 months after contract signing, that steel deliveries commence 3 months after contract signing, and that delivery takes place after 16 months. The cash flows will then be as shown in Table 31.

Table 31

Event	Time from signing	Owners capital payments	Loan drawdown	Drawdowns interest on loan	Owners N.C.F.
Signing of contract	0 months	5	—	—	5
1,000 tons steel delivered	3 months	5	—	—	5
Commencement of steel fab.	6 months	—	15	—	—
25% steel erected	8 months	—	20	—	—
75% steel erected	11 months	3	17	—	3
—	12 months	—	—	1.17	1.17
Steel erection completed	13 months	3	12	—	3
Launch	14 months	1	4	—	1
Delivery	16 months	3	12	2.23	5.23
		20	80	3.40	23.40

The NPV at 15 per cent as 21.254 per cent and at 20 per cent is 20.663 per cent. But in this case 'year 0' is 16 months prior to delivery rather than 24 months in the other cases. To make a true comparison we must therefore discount these values back a further 8 months giving figures of 19.363 (15 per cent) and 18.298 (20 per cent). Thus the improvement compared to the original situation is a reduction in the NPV's of about 1.75 equivalent to increasing the IRR in the particular case by about 0.9 per cent from 17.3 per cent to 18.2 per cent. This improvement, and the one obtained by reducing the building time only, show the importance of an early delivery and short building period on the owners financial analysis — the saving in interest charges also reducing his cash requirements during the building period. The builder will also be able to make a smaller allowance for inflation in his price. The actual figures will of course vary from case to case but the foregoing provide a guide.

The circumstances surrounding any particular contract may well in themselves present opportunities for improvement.

Where cheap credit is available it will also be advantageous to roll up the interest during construction unless this attracts a higher rate of interest which is equivalent to commercial funding (if available) for the rolled up interest — a matter which can be established by calculation. Other points regarding cheap loans are firstly that clearly any moratoria on repayments must increase the attractiveness of the package to the owner, unless the moratoria involves the addition of expensive commercial money. A further point on loans is that some forms of security do not cost the owner anything (except perhaps legal fees) — e.g. mortgages, charter party assignments — whereas other forms have a definite cost such as the bank guarantees. The builder should therefore do whatever he can to try and get the former type of security accepted.

Irrespective of the form of financial appraisal used by the owner, clearly the best way of improving the financial attractiveness of a new building contract for him will be a reduction in his own cash outgoings at as early a date as possible to improve his cash flow and the financial analysis. Just how effective such a reduction can be is demonstrated by using the calculation shown in Table 14 (p.45) and in Diagram 6. Suppose we reduce the owners cash expenditure in year 2 from 2.00 to 1.00 and add this sum to the contract price. The IRR will change from 17.27 per cent to 17.35 per cent — admittedly only a small change but nevertheless a move in the right direction which could be significant in a marginal situation also the owner's cash requirements up to delivery are reduced by 3.7 per cent. Of course if the builder can absorb this sum in the contract without addition to the price the improvement to the IRR would be much greater — to 17.69 per cent. This is more effective in improving the contract than a similar drop in price which will lower the amount of cheap credit being utilised, and which benefits the owner much less in terms of reducing his cash flow. In fact a reduction of 1.00 in the contract price would improve the IRR to 17.63 per cent but the cash flow improvement would only be 1 per cent against 3.7 per cent. Because the reduction in the owner's cash expenditure is therefore more attractive to him than a reduction in the price this can be carried a stage further (after all possible owners costs have been added to the contract) by the builder arranging to pay the owner substantial sums for some service or other during the building of the ship. In the case of a specialised vessel it could be for consultancy services in the design, or for any vessel it might be for feedback on performance after delivery, or some other arrangement appropriate to the situation.

Many shipbuilders now offer a choice of currencies for the contract and for the credit. There are three ways in which an owner can take advantage of the possibility of contracting in a currency other than his own accounting currency. Firstly it may enable him to 'match' the currency in which the value of his income is determined with that of his loan payments, as described in Chapter 7. Secondly the forward rates for an alternative currency may be such that by entering into forward contracts the vessel can be purchased for substantially less than in the original currency as described on p.54 without any element of speculation or risk. Thirdly there may be the possibility of successful speculation without undue risk by contracting and taking a loan in a weak currency which will devalue thus reducing the cost of the owners payments in his own currency.

Beyond these possibilities there may be further opportunities if the project is considered on an international basis. It may be, for example, that subsidies available in a particular country make it well worth while setting up a subsidiary in that country to own the vessel and charter it to the parent — indeed some owners have done extremely well by exploiting such possibilities to the full.

4. Valuation of a Company

This section is not intended as a full guide to valuation for takeover bids and mergers, but rather to draw attention to important points that arise if ships are acquired by the purchase of owning companies such as the "one-ship" companies already referred to. As a consequence of the enormous changes in ship values as a result of fluctuations in the charter market (p.116) shipping companies usually make no attempt to show their ships at valuation in their accounts because of the misleading effects this would have. Rather the ships are always shown at book value.

If the possibility of buying a shipping company arises, the first action must be to obtain a market valuation of the vessels owned. However, even if the company owns only one ship and has no other assets and liabilities, there is an important difference between the purchase of the ship and the purchase of the owning company. This difference is in the capital allowances. If the ship is purchased the new owner will obtain full capital allowances based on the purchase price, whereas if the owning company is purchased, only those allowances remaining, if any, will be acquired. Depending upon the level of initial allowances and rates of tax, this could make an acquisition through an owning company initially twice as expensive as direct purchase of the ship (e.g. UK pre 1984).

On the other hand there may be a favourable "newbuilding" loan on the vessel, which represents far more advantageous financing than could be obtained for a purchase of a similar vessel. Provided therefore the terms of the loan are not affected by the change of ownership of the company (e.g. security arrangements) this may offset the disadvantage re capital allowances.

Appendix 1

Appraisal of Shipping Projects — Further Considerations

In Chapter 6 we have described the classic DCF method of analysis and the usual manner of its application to shipping projects. However in the Author's view more useful results can be obtained through a modification to the usual methods in order to take account of the effects of the large loan funding on the calculated rates of return.

When large loans (e.g. 80 per cent or more of the total capital invested) are included in the cash flows in the manner already described (e.g. Table 14) the profits of the project are related only to the small part of the capital invested remaining to be provided by the investing company. Consequently the returns on the capital invested become very susceptible to small changes in the cash flows. This can be illustrated by considering a hypothetical investment situation such as that already described and detailed in Table 15. On the assumption of an annual income for capital servicing of 18 for ten years, and nil residual for the vessels after this time we can calculate the IRR's as follows:

Table 32

Vessel	IRR
A	27.4%
B	23.2%
C	17.7%

On the basis of these figures it might be felt that investment in Vessel A showed a good margin to cover the risks of the project; we know that the only difference between A and B is a loan at 14 per cent pa, — assuming this is a secured bank loan we can reasonably expect the company would require a higher minimum rate of return than this on its own money and would therefore rule out Vessel B if it followed a strict financial basis (i.e. incremental analysis (p.43) would show 14 per cent on the increment, less than required). However suppose the rate of hire was a little less than assumed such that the income left for capital servicing fell by 20 per cent. The new IRR's would be:

Table 33

Vessel	IRR
A	5.7%
B	7.5%
C	10.3%

This table shows a very different story. All the IRR's are now in fact below our assumed minimum for the company, but had they invested in Vessel A they would have fared worse than if they had proceeded with either B and C. The reader will notice how the highest IRR has become the lowest and vice versa and that most violent changes have occurred in the vessel with the highest proportion of loan funding.

This result is to be expected because in the case of Vessel A for example all the risks and profits of the project are borne by only 10 per cent of the capital expenditure which must bring into question the validity of including the cash flows relating to the loan capital in the DCF calculation. Clearly if they are included in such a case, the IRR calculation is of use only if the cash flows can be virtually guaranteed.

Academics argue that all the cash flows relating to a project should be included in the DCF calculation including loan cash flows even in the type of cases considered above. However this overlooks the real purpose of the IRR calculation — to give an indication of the profitability of a project which the directors can compare with the risks.

Because of the distortions produced when the loan cash flows are included some companies now ignore loans in their calculations. If we do this the IRR's for the three alternatives in Table 15 become:

Table 34

	IRR ignoring loans	
Vessel	income 18 pa	income 14.4 pa
A	12.4%	7.25%
B	12.4%	7.25%
C	16.6%	10.95%

A company basing its decisions on this analysis should choose Vessel C if we assume a minimum acceptable return of say 15 per cent, and an income of 18 pa, in contrast to the choice presented by Table 32 where the choice on pure financial grounds would be for Vessel A. The calculations producing Table 32 can be criticised in a risk situation on the grounds that the profits are related to only a small part of the capital at risk, while the calculations leading to Table 34 ignore the benefits of the low rate of interest on the shipbuilding loan. A more realistic approach would appear to be to base our calculations on the total capital at risk but also to take credit for the low interest rate on the loan.

Suppose that instead of regarding the loan as being a part of the project only we regarded it as part of the owner's capital in the same way as we would if it had been raised as, for example, debenture stock, for the purpose of financing the project, being secured on the new vessel and on other assets (exactly as the subsidised loan might be secured). In this case the loan or loans would be included in the calculation of the cost of the capital but in a large company the loans might have little effect on the overall cost — also the calculation for the cost of capital for a company is usually complicated if indeed it is possible to make a meaningful calculation at all.

We must therefore consider alternative means for evaluating the benefit of a cheap loan. Two possibilities suggest themselves — firstly a comparison with a commercial loan rate and secondly a comparison with the return on a risk-free investment. The former would compare the loan to an alternative source of funds which may be a relatively expensive source and therefore give an exaggerated apparent benefit from the cheap loan. The second possibility which shows the benefit that would derive from the investment of the loan funds in a risk-free investment, if a loan is available at a rate below the gilt-edged rate there is clearly a benefit in that the loan funds could in theory be invested elsewhere, and contribute a return to the venture without risk. Conversely if funds are borrowed at a rate above 'gilt-edged' then there must be a penalty (because the difference is being covered twice — by both the lender (e.g. the bank) and the investor of the borrowed funds). Particularly if the judgement as to minimum acceptable IRR is based on comparison with a risk free investment this second method is to be preferred.

The differences between the rates of interest payable on loans provided for a project and the gilt-edged rate should be taken into account by comparing the cash flows of loan drawdowns and repayments (including interest) with the return that could be earned if sums equal to the loan were invested in gilt-edged stock — the capital invested at any time equalling the loan outstanding at that time.

The gilt-edged redemption yields which could be obtained will depend upon the type of stock in which the hypothetical investment was made. The attraction of capital profits to many investors lowers the redemption yield on low interest stocks, but ignoring the tax situation it would be reasonable to consider the highest redemption yield available for the appropriate period.

The British Financial Times publishes a table showing the average gross redemption yields for British Government stocks in three groups according to the level of the coupon (i.e. the 'face value' of the interest paid on the stock), with each group divided in turn into three according to the life of the stock before redemption. For shipping projects the 5 year and 15 year lives are of interest and the difference in yield is usually small (e.g. ¼ per cent). Certainly therefore for preliminary analysis a flat rate irrespective of maturity can be used, and in practice a company could itself determine from time to time the rate to be used in project analysis.

The following table shows the comparison between the loan repayments in the case of Vessel B and the income which could be derived if a similar sum was invested in stock having both an interest yield and a redemption yield of 12 per cent.

Table 35

Year	Loan			Gilt-Edged			
	Capital	Interest	Total	Capital	Interest	Total	Difference
0	80	—	80	80	—	80	—
1	(9.41)	(5.82)	(15.23)	(9.41)	(9.32)	(18.73)	3.50
2	(9.41)	(5.12)	(14.53)	(9.41)	(8.20)	(17.61)	3.08
3	(9.41)	(4.41)	(13.82)	(9.41)	(7.06)	(16.47)	2.65
4	(9.41)	(3.71)	(13.12)	(9.41)	(5.93)	(15.34)	2.22
5	(9.42)	(3.00)	(12.41)	(9.42)	(4.80)	(14.22)	1.81
6	(9.41)	(2.24)	(11.70)	(9.41)	(3.67)	(13.08)	1.38
7	(9.41)	(1.59)	(11.00)	(9.41)	(2.54)	(11.95)	0.95
8	(9.41)	(0.88)	(10.29)	(9.41)	(1.41)	(10.82)	0.53
9	(4.71)	(0.18)	(4.89)	(4.71)	(0.28)	(4.99)	0.10

The difference in the last column thus measure the benefit that would be derived from investment of the loan in a risk free investment; these differences are therefore added to the cash flows calculated ignoring the loan to give the overall net cash flow for the DCF calculations. Instead of the calculations shown in Table 14 we then have the following:—

Table 36

Year	Capital expenditure on contract	Othei non-operating expenditure & income	Loan benefit	Net charter income	Notional N.C.F.
0	(5.00)	(1.00)	—	—	(6.00)
1	(40.00)	—	—	—	(40.00)
2	(55.00)	(2.00)	—	—	(57.00)
3	—	—	3.50	16.50	20.00
4	—	—	3.08	16.50	19.58
5	—	—	2.65	16.50	19.15
6	—	—	2.22	16.50	18.72
7	—	—	1.81	16.50	18.31
8	—	—	1.38	16.50	17.88
9	—	—	0.95	16.50	17.45
10	—	—	0.53	16.50	17.03
11	—	—	0.10	16.50	16.60
12	—	—	—	16.50	16.50
13	—	—	—	16.50	16.50
14	—	—	—	16.50	16.50
15	—	—	—	16.50	16.50
16	—	—	—	16.50	16.50
17	5.00	—	—	16.50	21.50
	(95.00)	(3.00)	16.22	247.50	165.72

The IRR for the above notional net cash flows is 14.4 per cent.

Reverting to the three vessels in Tables 15, 32, 33, and 34, if we apply this method to them with the risk-free rate quoted above of 12 per cent we have the following results for comparison (Table 37):

The IRR's on Vessels A and B have improved compared to Table 34 because their loans are below the 'gilt-edge' rate; the IRR on Vessel C is reduced because its loan is above that rate.

Table 37

Vessel	Notional IRR	
	income 18 pa	income 14.4 pa
A	15.05%	9.93%
B	15.21%	10.10%
C	16.20%	10.52%

To further illustrate the problems of handling analysis when large loans are involved we can divide the cash flows into the outgoing cash flows covering the purchase of the vessel and the incoming cash flow from trading activities. The NPV's for these cash flows can be plotted for a range of discount rates.

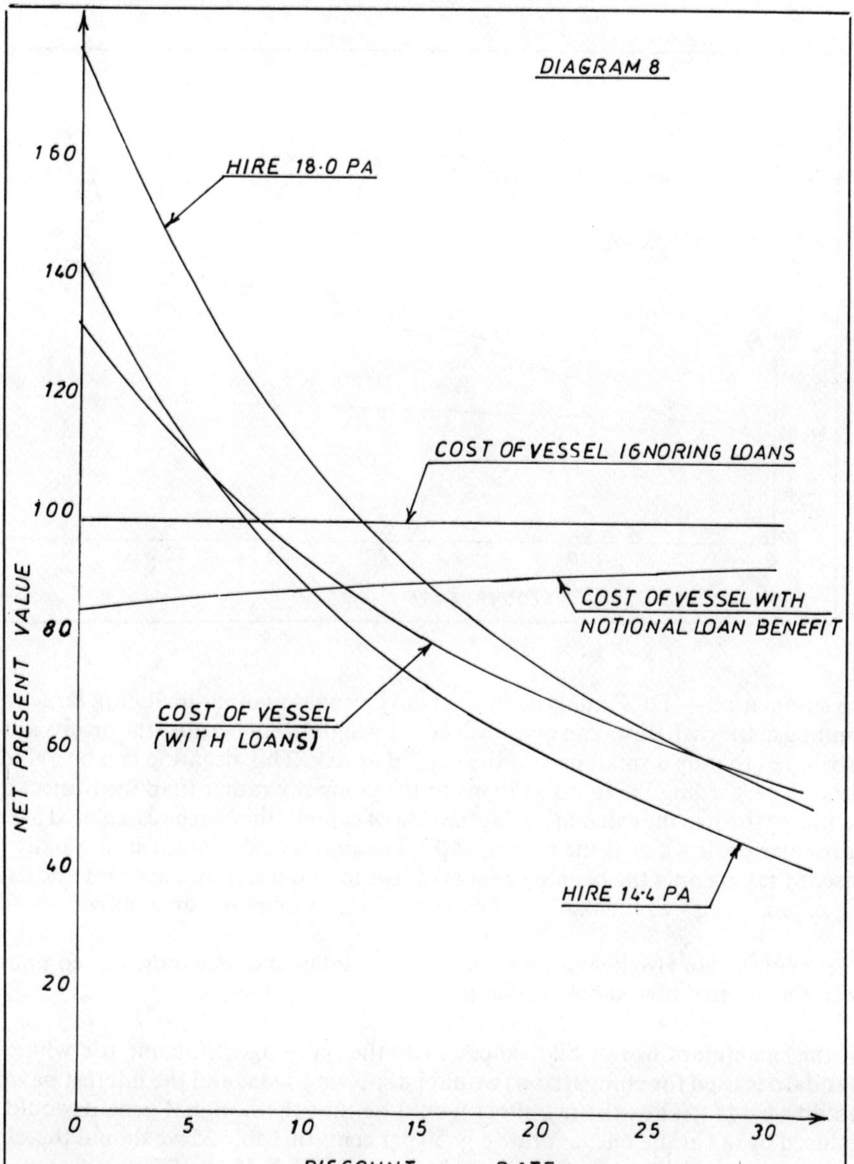

This diagram shows Vessel A and illustrates how a small change in income will cause a major change in the IRR (the point where the NPV's of outgoing and incoming cash flow are equal when the loan is included and the profits are calculated on only a small part of the capital at risk). The NPV ignoring the loan is shown as is the NPV modified to take account of the benefit of the loan interest as suggested.

We should now refer again to Diagram 7 (page 49). This diagram shows the NPV's of the three vessels calculated in the conventional way including the loan cash flows. Notice how the choice of discount rate will change the apparent relative merits of the three alternatives, thus showing the dangers of comparing alternative offers on such a basis. The Author prefers to take only the benefits of the low interest rates into account when they occur, giving the NPV's shown in the following diagram.

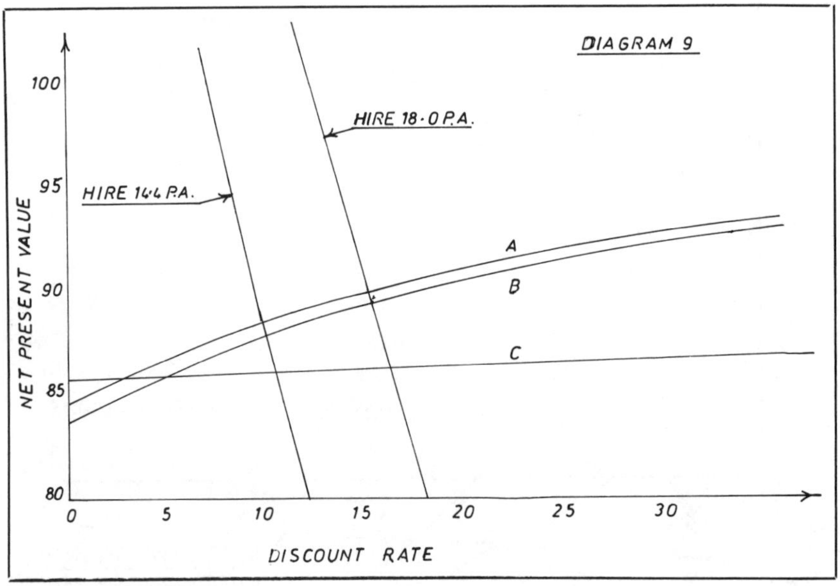

To summarise — DCF analysis in the conventional manner including large loan funding in the cash flows can give misleadingly high IRR's because the profits are all attributed to only a small part of the capital at risk. This situation can be avoided either by regarding the loans as loans to the company rather than the project and including them in the calculation for the cost of capital (this can be calculated for the particular project), or if the cost of capital cannot be calculated, as is usually the case, by taking only the benefit (or cost) of the loan interest into account and using the capital at risk as the basis of the calculations as demonstrated above.

This method can also be used when taxation is taken into account, and compared with the 'no-tax' case shown in Table 36.

In the 'maximum use of allowances case' the 'gilt-edged' income (or whatever standard is used for comparison) would be subject to tax, and the interest payable would save tax. Thus the net effect would be that the notional benefit would be reduced by tax at the relevant rate e.g. 50 per cent. In Table 25 we should therefore add as notional benefit the benefits calculated in Table 35 less 50 per cent tax, e.g.

Table 38

Year	Net cash flow after tax (Table 25)	Notional loan benefit less 50% tax	Notional cash flow after tax
0	(6.00)	—	(6.00)
1	(37.00)	—	(37.00)
2	(37.00)	—	(37.00)
3	45.00	1.75	46.75
4	8.25	1.54	9.79
5	8.25	1.32	9.57
6	8.25	1.11	9.36
7	8.25	0.91	9.16
8	8.25	0.69	8.94
9	8.25	0.47	8.72
10	8.25	0.27	8.52
11	8.25	0.05	8.30
12	8.25	—	8.25
13	8.25	—	8.25
14	8.25	—	8.25
15	8.25	—	8.25
16	8.25	—	8.25
17	13.25	—	13.25
18	(10.75)	—	(10.75)
	74.75	8.11	82.86

IRR = 14.17%

In the case where the project is regarded as an isolated one for tax purposes, the benefit should be shown gross until a tax paying position is reached, and net of tax thereafter. It could be argued that as we are treating the benefit as income, the allowances should also be used against it, thus advancing the point at which tax is first paid, but as the benefit is notional this would be an unnecessary complication.

Appendix 2

Calculation of single rate of interest on combination of loans

If we wish to extend the period of a loan by taking out an additional loan at a different rate of interest, combining the loans in such a way that the combined repayments with interest take the form of equal repayments of capital together with interest on the capital outstanding (as shown in Chapter 4) then the rate of interest that must be applied to the combination can be calculated in the following way.

The present value of loan capital repayments plus interest payments will equal the sum drawndown initially if discounted at the rate of interest charged, i.e. the capital repayments and interest payments on a loan of 100 on which interest is paid at 15 per cent will have a present value of 100 if discounted at 15 per cent.

Consideration of the example in Chapter 4 (Table 5) shows that for each period:-

Total payments on combined loan	=	Total subsidised loan payments payments drawdown on commercial loan

Therefore if we discount,

PV payments on combined loan	=	PV subsidised loan payments ± PV payments/drawdowns on commercial loan

But if we discount at rate of interest on commercial loan, PV of this loan will equal initial drawdown, which in example = 0.

Therefore discounting at a nominal 15 per cent pa (7.5 per cent for a six month period.)

PV capital payments on combined loan + PV interest payments on combined loan	=	PV subsidised loan payments.
Now PV subsidised loan payments	=	62.20 by calculation.
PV of combined loan capital payments	=	3.3333 × PV of 1 per period at 7.5 per cent for 24 periods.
	=	36.61

Therefore PV of interest payments on combined loan if at 15 per cent per annum would be $= 80 - 36.61$
$= 43.39$

But actual PV of interest payments $= 62.20 - 36.61$
$= 25.59$

Therefore rate of interest on combined loan $= 15 \times \frac{25.59}{43.39}$
$= 8.85$ per cent.

See Table 5 which acts as proof.

This method can also be used where the commercial loan is to be used to give a combined loan of say 90 per cent, i.e. 10 per cent of the total cost will be provided by a commercial loan. In the case the PV of the commercial loan repayments and interest would be = 10, and if p = rate of interest on combined loan,

$$62.20 + 10 = 36.61 + 43.39\ p/15$$

from which $p = 12.30$ per cent

If the commercial loan is required to give a moratorium on capital repayments the calculation is still performed in exactly the same way but the PV of the combined loan will reflect the moratorium, e.g.

Suppose basic loan is 80 per cent over $8\frac{1}{2}$ years at $7\frac{1}{2}$ per annum;
Commercial loan as required over 12 years at 15 per cent per annum (no initial drawdown)

Combined loan to be repaid in equal semi annual instalments of capital, the first such payment to be 30 months after delivery of vessel and the last 12 years after delivery. Interest to be paid semi annually in arrears with the first payment six months after delivery.

In this case we have:

PV of basic loan repayments etc. as before = 62.1991
PV of commercial loan, as before = 0
PV of combined loan: capital repayments 4.00 per half year, 0 for first four half years
$PV = 4\ (10.9830 - 3.3493) = 30.5347$
PV of interest payments = 80 – 30.5347 = 49.4653
Then if actual rate on combined loan = p per cent pa
$62.1991 = 30.5347 + 49.4653\ p/15$
from which p = 9.602 per cent

See Table 39 which confirms that this is the correct rate.

Table 39

	Combined loan			Commercial loan			
Year	Basic loan total repayments	Capital repayments	Interest payments at 9.602% pa	Total payments	Capital (drawdown) repayment	Interest capital at 15% pa	Loan o/s
0							
1	7.706	—	3.841	3.841	(3.865)	—	3.865
	7.530	—	3.841	3.841	(3.689)	0.290	7.844
2	7.353	—	3.841	3.841	(3.512)	0.588	11.944
	7.177	—	3.841	3.841	(3.336)	0.896	16.176
3	7.000	4.000	3.841	7.841	0.841	1.213	16.548
	6.824	4.000	3.649	7.649	0.825	1.241	16.964
4	6.647	4.000	3.457	7.457	0.810	1.272	17.426
	6.471	4.000	3.265	7.265	0.794	1.307	17.939
5	6.294	4.000	3.073	7.073	0.779	1.345	18.505
	6.118	4.000	2.881	6.881	0.763	1.388	19.130
6	5.941	4.000	2.689	6.689	0.748	1.435	19.187
	5.765	4.000	2.497	6.497	0.732	1.486	20.571
7	5.588	4.000	2.304	6.304	0.716	1.543	21.398
	5.412	4.000	2.112	6.112	0.700	1.605	22.303
8	5.235	4.000	1.920	5.920	0.685	1.673	23.291
	5.059	4.000	1.728	5.728	0.669	1.747	24.369
9	4.882	4.000	1.536	5.536	0.654	1.828	25.542
	—	4.000	1.344	5.344	5.344	1.916	22.114
10	—	4.000	1.152	5.152	5.152	1.659	18.620
	—	4.000	0.960	4.960	4.960	1.397	15.057
11	—	4.000	0.768	4.768	4.768	1.129	11.418
	—	4.000	0.576	4.576	4.576	0.856	7.698
12	—	4.000	0.384	4.384	4.384	0.577	3.892
	—	4.000	0.192	4.192	4.192	0.291	0.008

Appendix 3

Ships and Mobile Offshore Installations

The distinction between ships and mobile offshore installations is important to the owner not only from a credit standpoint but also for purposes of taxation and legal requirements. In the UK the distinction is as follows:—

A mobile offshore installation is defined as "any installation intended for underwater exploitation of mineral resources or exploration with a view to such exploitation and which can move by water from place to place without major dismantling or modification, whether or not it has its own motive power". The definition of a ship is not as precise. The Merchant Shipping Act, 1894, (s.742) defines a ship as a vessel of any description used in navigation and not propelled by oars — this definition has been further clarified by case law since 1894.

Ships thus include supply vessels, support craft, diving vessels, etc. Note that the shape of the vessel is immaterial, i.e. there is no distinction between mono-hulls and semi-submersibles. Mobile offshore installations would include floating production platforms, jack up rigs, etc.

There are some vessels which meet both definitions, in particular drilling vessels. For credit purposes the Department of Industry (DOI) has classed drilling vessels as mobile offshore installations in accordance with an EEC directive although it is believed 'ship' credit terms have nevertheless been offered. (In some countries outside the EEC ship terms are readily available for such vessels.)

For UK Corporation Tax purposes, until 1984 ships uniquely enjoyed 'free depreciation' in that the first year allowances could be used as and when the owner wishes. A mobile offshore installation on the other hand, qualified (until 1984) for a 100 per cent first year allowance but any unused allowances after the first year could only be utilised at the rate of 25 per cent each year of the diminishing balance. Thus a drilling rig enjoyed free depreciation — a floating production platform did not. From 1986 however both will receive the same treatment, namely 25 per cent writing down allowances — no initial allowances.

On legal grounds, a ship is required to comply with the Department of Trade rules and regulations, which are based on international agreements for ships. A mobile offshore installation that is not legally a ship does not have to comply with these rules (but will have to comply with the Department of Energy rules and regulations).

Appendix 4

UK Government Credit Schemes

Home owners scheme

Under Section 10 of the Industry Act, 1972 the Secretary of State for Industry may guarantee the repayment of loans to UK bodies (together with the repayment of interest thereon) for the purpose of financing the construction to their order in the UK of new ships and mobile offshore installations. It should be noted that the Act does not require the Secretary of State to give guarantees but rather it gives him the opportunity to do so — nevertheless so far as is known there has never been a refusal to provide credit where the various criteria required have been met. The Act is in no way restricted as to place of build except that construction must take place within the UK.

The loans are made by the owner's bankers to the owner, i.e. buyer's credit. Under the arrangements that have been made the loans are made to the purchaser at a fixed rate of interest, (the difference between this (low) rate and the commercial rate being paid to the bank by the government).

The Department of Industry (DOI) have appointed the Ship Mortgage Finance Company (SMFC) to conduct all negotiations on their behalf, to assess the credit-worthiness of the borrower, to advise on the security required, to consider the documentation, etc.

It is understood however that in practice the SMFC deal principally with the security arrangements, etc. and refer matters such as rulings on foreign content to the DOI.

To be eligible for a loan under this scheme, the owner must be a person resident in, or a company incorporated under the law of any part of the UK, Isle of Man, or the Channel Islands. There is no objection to a company thus incorporated being controlled by an overseas organisation, but it must be an operating company and not just a brass plate.

The application for the loan is made by the owner to SMFC, having first obtained an undertaking from his bankers that the loan will be made available by them provided the guarantee is obtained. The application is then considered by SMFC and their advice passed to the DOI who then notify the owner whether or not the application has been approved in principle and, if so, on what terms. If the owner then wishes to proceed SMFC draw up a Financial Agreement, a Deed of Covenant, and an Owner's Agreement dealing with guarantee fees, etc.

It is important that the application for the loan is made prior to any work starting on the ship/installation or the full credit may not be made available.

Applications are considered at monthly meetings of an Advisory Committee at SMFC who make their recommendations to the DOI. Applications are required at least two weeks before such a meeting; the total time between submission of all necessary documents to SMFC and the Owner being notified of the DOI's decision is therefore usually in the region of four to eight weeks.

Loans are normally in Sterling in association with Sterling building contracts, but it is now possible for US Dollar loans to be arranged in association with building contracts in US Dollars.

The terms currently available (1984) for ships are for loans amounting to 80 per cent of the contract price of the vessel repayable over 8½ years, with interest charged at 7½ per cent pa. The repayment of capital is made in equal instalments at six monthly intervals, commencing six months after delivery to the purchaser. Interest payments are made at the same time.

For loans in respect of mobile offshore installations the maximum size of loan is 85 per cent of the contract price repayable over 5 years, with interest charged at the rate applying to buyer credit guaranteed by ECGD for exports other than ships.

The loan can be drawndown during the construction period, but the first 10 per cent of the contract price must be paid by the owner before drawing on the loan. During the construction period the loan drawdowns are limited such that at no time the ratio of loan drawdown to owners cash payments exceeds 85 : 15 in the case of 85 per cent loan or 80 : 20 in the case of an 80 per cent loan.

It is possible to 'roll-up' the interest payable on the loan drawdown during the construction period and to include this in the loan to be repaid after delivery. However if this is done the rate of interest on the loan (from the first drawdown) is increased from 7½ per cent to 8 per cent.

Security for the loan will usually take the form of a charge on the vessel or installation during construction and a first mortgage after delivery until the loan is repaid. In addition further security is usually required such as a mortgage on further assets, assignment of a charter party, or a guarantee from a bank or other third party.

The security e.g. the first mortgage, is given by the owner to the lending bank, but the extent of the security required is determined by SMFC. In the case of a ship of a type for which there is an established sale and purchase market, e.g. a bulk carrier, SMFC will usually accept the first mortgage as covering 50 per cent of the contract price. In the case where the ship or mobile offshore installation is of a more specialised nature, the value placed on the first mortgage will be somewhat less, e.g. 40 per cent.

It is a condition of the loan guarantee that the ship or installation is constructed and equipped throughout with goods and materials of UK origin.

Nevertheless it is recognised that his may not always be possible and foreign goods and materials will be allowed if similar products are not available from UK sources on competitive terms. If it is proposed to include any foreign goods or materials the DOI must be advised with the details of the reasons for their use and provided the DOI agree the maximum loan is not affected until the total value of such goods and/or materials reaches the value of the cash payments by the owner i.e. 15 per cent or 20 per cent, after which point the maximum loan is reduced such that the payment for the foreign content is covered by the owner's cash payments.

If the DOI are not satisfied that the use of foreign goods or materials is justified then the maximum loan will be reduced pro-rata.

Upon the signing of the Owner's and the Financial Agreements, the Owner is required to pay a guarantee fee to the DOI of 1/3 of 1 per cent of the sum of the loan principal and post delivery interest. In the case of a ship with an 80 per cent loan over 8½ years at 7½ per cent interest this would amount to about 0.36 per cent of the contract price.

At the same time the Owner will be required to pay fees to his bankers, the level of which can now be negotiated. Formerly there was a fixed fee of 1 per cent of the principal amount of the loan (i.e. 0.80 per cent of the contract price in the case above), with the provision for a possible small additional charge.

In addition, upon indicating to SMFC that he wishes to proceed, (having accepted the terms offered in principle), the Owner is liable for SMFC's legal costs in addition to his own. These costs vary considerably but could amount to about £100,000 in total.

Export Credit Scheme

Government supported export credit schemes are administered by the Export Credits Guarantee Department (ECGD) — a government department responsible to the Secretary of State for Trade — under powers derived from the Export Guarantees and Overseas Investment Act, 1978, although ECGD was originally set up in 1919.

The scheme appropriate to the export of ships and mobile offshore installations is the buyer's credit scheme, under which ECGD give the lender a guarantee against non-repayment of both principal and interest, whether due to default of the buyer or restrictions (e.g. exchange control) imposed by the buyer's country, and enable the lender to offer favourable rates of interest.

To be eligible for export credit the owner must be resident outside the UK, (including Isle of Man and the Channel Islands). The intended place of use of the ship or mobile offshore installation is not considered relevant.

Applications for export credit is made to ECGD by the builder who must nominate a UK registered bank as the lender. ECGD then assess the risks involved, and, if prepared to give support, will issue a letter of intent giving the terms upon which support will be given. ECGD's assessment and decision can usually be obtained in 6-8 weeks, but can be obtained much quicker if good reason for a quick decision can be shown, or if the buyer is already known to ECGD. The timing of their decision is not determined by particular meetings, etc.

Upon being advised that the builder and owner wish to proceed on the terms set out, ECGD draw up the necessary documents including the financial agreements between the owner and the lending bank, and the agreement with the builder dealing with the premium and recourse. The arrangements of the agreements for ECGD cover are illustrated diagrammatically in Diagram 10 which also shows the UK owners scheme for comparison.

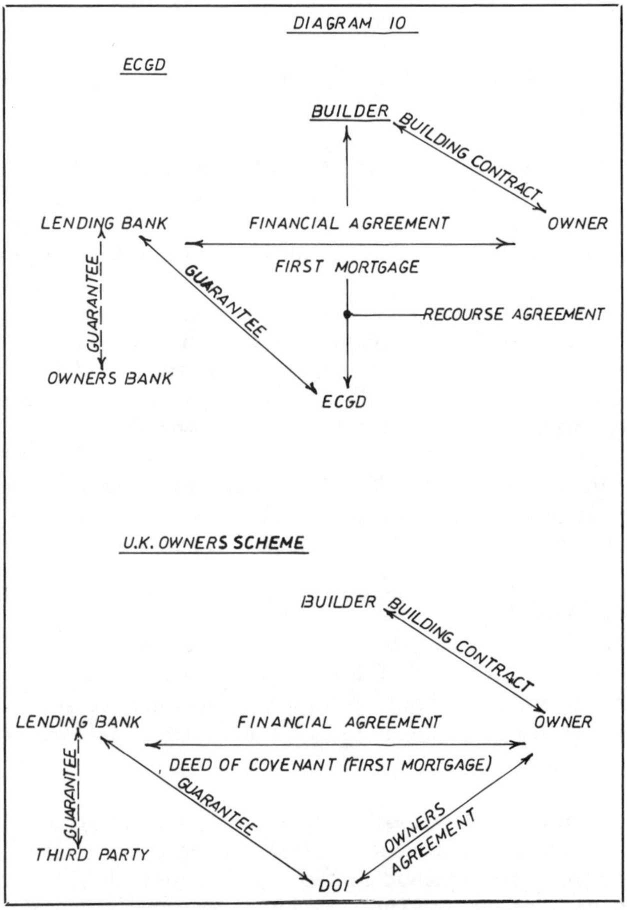

The currency of the loan is usually US Dollars or Sterling although guarantees can be given in respect of some other currencies such as Deutchmarks.

The terms available through ECGD for ships are the same as those for UK owners, i.e. 80 per cent of the contract price repayable over 8½ years from delivery, with interest at 7½ per cent.

The loan can be drawndown during the construction period in exactly the same way as for UK owners, except that ECGD require only 5 per cent initial cash payment by the owner (rather than the 10 per cent required from a UK owner) before loan drawdowns can begin.

As with the UK scheme, it is now possible to roll up the interest on the loan drawndown during the construction and add this to the loan principal at delivery. Such an arrangement would result in an increase of about ½ per cent on the rate of interest and would increase the premium payable to ECGD.

Another facility offered by ECGD is cover against charging currency exchange rates between the time of tender and contract. It is possible, by paying a premium to 'freeze' the forward rates for the currency of the contract at the time of tendering. If the bid is successful and the contract is concluded a further premium is payable and ECGD then take any profit or loss that may arise as a result of charges in forward rates since the time of tendering.

There are differences in the security sought by ECGD and SMFC deriving partly from the fact that the owners' bankers are usually actively involved in a loan to a UK owner whereas, they are not involved in the case of export credit.

The preferred security for export credit is a bank guarantee for the principal and interest, but in practice the more usual security is a first mortgage together with a bank guarantee covering a number of initial payments. This guarantee might come from the owner's bankers. ECGD will value a first mortgage on a 'marketable' ship as 55 per cent of the contract price, (cf 50 per cent by SMFC), but this would reduce as the ship, (or mobile offshore installation), became of a more specialised nature. This would be a matter for negotiation.

ECGD are reluctant to consider assignment of a charter as part of the security package as they find it difficult on legal grounds and have had unfortunate experience of this form of security.

In addition to the security requirements outlined above ECGD have recourse to the builder, but only if, when the borrower is in default under the financial agreement, the builder is in default under the building contract. Such recourse is usually limited to 10 per cent of the loan; the builder is normally released from this obligation 12 months after delivery provided all guarantees have then expired and there is no loan default or other difficulty. It is not usually considered necessary to show such a recourse on the builders balance sheet. There is no similar recourse arrangement in the UK owners scheme.

Any non UK supplied goods or materials incorporated in the ship or mobile offshore installation must be declared to ECGD. The effect of incorporating such goods and materials on the loan is similar to that described for the UK owners scheme.

The premium payable to ECGD in respect of their guarantee is paid by the builder upon the signing of the Recourse Agreement between the builder and ECGD. This signing is usually followed immediately, (i.e. the same day), by the signing of the ECGD Guarantee Agreement with the lending bank and the Financial Agreement together with associated security agreements. It is quite common for the Building Contract to have been signed some weeks before, subject to the signing of the Financial Agreement within a given period.

The amount of the premium is determined by a number of factors, principally the contract price, the country of residence of the owner, and the credit worthiness of the owner. It could be of the order of 3 per cent of the loan principal and interest, e.g. 3.9 per cent of the contract price. It is thus very much greater than the guarantee fee charged under the UK owners scheme.

Because the premium can thus be a substantial sum, ECGD are prepared to accept it in a series of instalments. They do however require a bank guarantee for the instalments, and limit the drawdowns to the same proportion of the loan as the proportion of the premium that has been paid.

The Financial Agreement, etc. are drawn up by ECGD's in-house solicitors and the builders legal charges are therefore limited to his own solicitors in connection with the building contract — similarly the owner has only his own solicitors to pay.

Bank commitment and negotiation fees would be the subject of negotiation between the owner and his bank.

Appendix 5

The Shipping Market Supply and Demand

An observer of the shipping industry might conclude that it was either enjoying boom conditions (occasionally) or was in a state of depression or slump (most of the time). In fact this is very much the case due to the inherent supply and demand characteristics of the industry giving rise to very large variations in prices paid for shipping space.

The supply and demand for most commodities are self-regulating, in that an increase in demand for a particular commodity leads to an increase in price which in turn has the effect of reducing the increased demand, (for example some purchasers cannot afford the higher price), and increasing the supply, (for example high cost producers enter the market).

In the case of shipping however the prices paid for shipping space — freight rates — are usually only a small fraction of the final price paid for the goods shipped. For example, in the case of petrol produced and sold in the UK from crude oil shipped from the Middle East, only 1 per cent of the selling price represents the shipping cost. Thus freight rates can increase several times over before having any effect on the demand. Furthermore the supply of shipping space, i.e. of ships, is relatively inflexible since even with spare capacity in the shipbuilding industry it will still take something like twelve to eighteen months from contract to delivery for a new vessel.

Thus it can be argued in a theoretical case that if there are 99 cargoes for 100 ships rates will fall until one owner decides that his losses with his vessel laid up out of service will be less than with it operating. Similarly if there are 101 cargoes for 100 ships rates will rise to very high levels before 1 cargo is no longer worth sending.

The actual fluctuation of freight rates in real terms over the period 1960-1982 is shown in diagram 11.

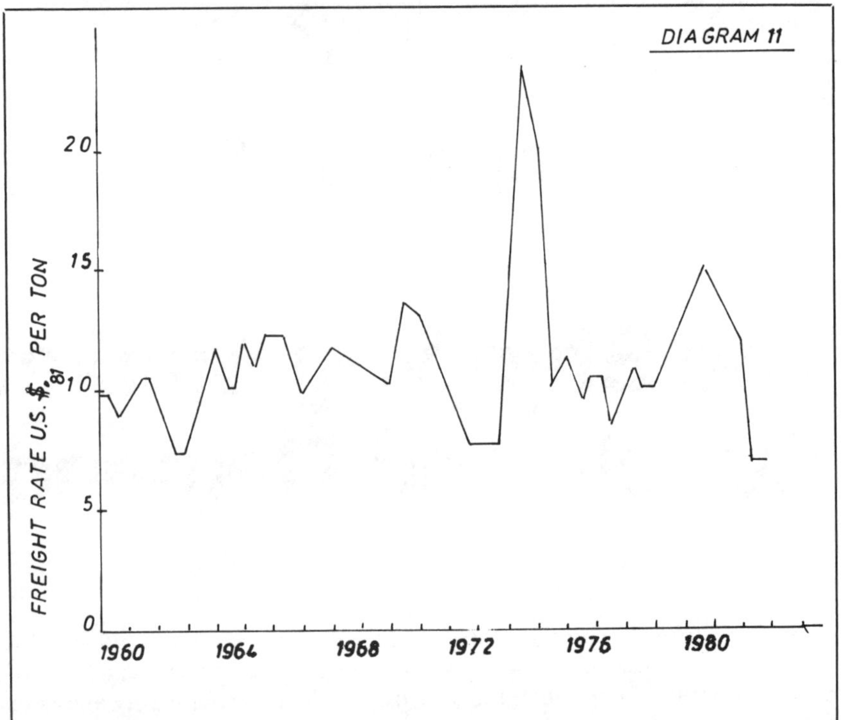

BASED ON G.C.B.S. TRAMP TIME CHARTER INDEX (GROUP 1)

A good practical example of the fluctuation of rates is shown by the market for semi-submersible drilling rigs for which there is no substitute for most of their tasks, and where the rate of hire has little effect on demand. This fluctuation is shown in diagram 12. It is worth noting that before rates started to rise in 1979 well over 90 per cent of the world fleet was operating, indeed only some 6 per cent were without work. Rates at that point were still at a level which was probably below operating costs for most owners, and which certainly made no contribution towards reimbursing the owner for his purchase of the rig.

Thus small changes in the supply or demand for vessels can have an effect on earnings out of all proportion to the size of the change. This makes the forecasting of periods of high earnings extremely difficult.

The forecasting of shipping demand is made all the more difficult because a part of it meets marginal requirements, for example, in prosperous times liner companies may charter in additional vessels to supplement their own. Most oil companies have their own fleet supplemented by chartered vessels for their marginal requirements.

Nevertheless shipowners are constantly trying to anticipate the market changes, and a number of organisations exist primarily to write and sell reports on market movements, and demand for crafts of different types.

One of the more important factors determining ship demand is the general level of world trade, and of industrial activity throughout the world.

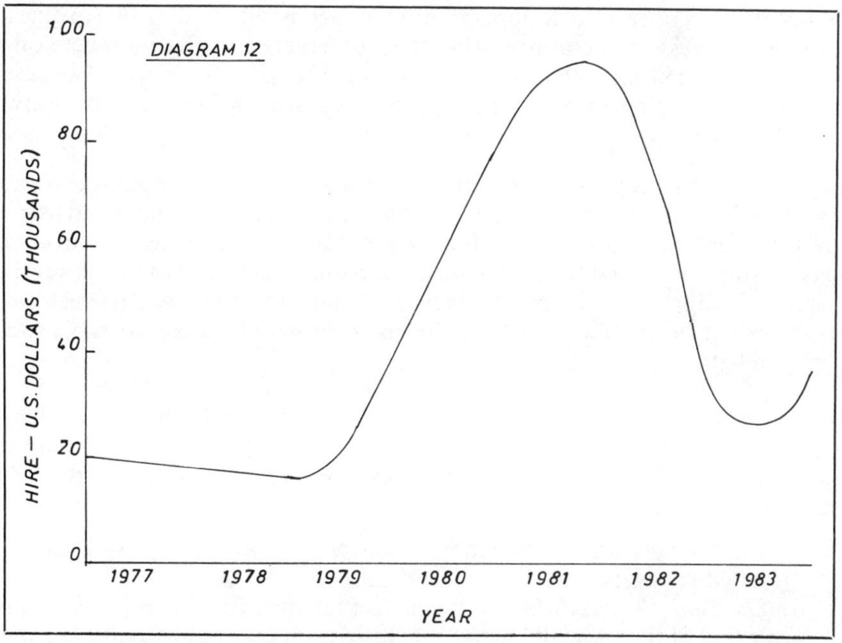

To some extent there may be different demand patterns for different types of craft. Specialised craft will certainly not necessarily follow the bulk shipping demand, except that if they are carrying materials for industry or products of industry their patterns of demand may be determined by similar factors. Other craft, for example semi-submersible drilling rigs, may have an almost completely different cycle of demand.

In the past quite separate situations for demand/supply of ships could exist between the oil tank ship and the dry bulk market. The use of surplus tankers to carry grain cargoes and more recently the development of OBO vessels able to carry either oil or dry bulk cargoes has resulted in a degree of interchangeability of ships between the two markets so that they now and in the future will probably follow similar patterns, quite apart from the fact that they are both influenced by the same basic factors of world trade levels, etc.

Employment

Ships may be employed by their owners in a number of ways. Some, will be used to carry their owner's goods. In other cases the shipowner will use his ships to provide a service to the world at large, operating the vessels on fixed routes at advertised intervals, and carrying probably many small consignments of goods for many different shippers. Such services, known as liner services, hence the passenger liners of the past, have in many cases been containerised in recent years, and now carry the goods in containers.

On some of the busier shipping routes a number of companies may each offer liner services. In most such cases the companies co-operate with one another and form a Conference. The Conference will co-ordinate the services and fix the rates to be charged for different goods, leaving the companies to compete with each other in terms of efficient operation. Shippers who ship their goods on the route in question exclusively by vessels operating within the Conference are given a rebate on the published rates, but this rebate is not paid until after a certain period has passed, thus discouraging shippers from using any service which might be set up in competition to the Conference lines.

The principal advantages to the Conference system are that the shipping companies are assured of a fair chance of profit while the stable rates and regular and frequent departures enable shippers to make forward contracts. The principal disadvantages to a large shipper are that he cannot use his bargaining power to obtain lower rates, the same rates apply to everyone, nor can he take advantage of cheap tramp tonnage during periods of depression without losing what may be a considerable sum in deferred rebates.

Other owners employ their ships carrying cargoes all over the world according to the cargoes available. This is tramp shipping. The arrangements for carrying such cargoes are embodied in agreements known as Charter Parties. There are three main types of Charter Party:

i) *The Voyage Charter*: which is a contract for the carriage of goods on a stated voyage;
ii) *The Time Charter*: which is a contract for the hire of the ship for a given period of time, usually with some restriction on trading limits, etc.
iii) *The Demise Time Charter or Bareboat Charter*: under which the charterer virtually becomes the owner for the period of the charter, the legal owner becoming little more than a financier. This charter is used in leasing operations.

The difference between these different types of charter can be seen from the following tables:—

Type of Charter	Owner provides and pays for:-	Charter provides and pays for:-
Voyage	Crew, repairs, insurance etc. Fuel, port charges	—
Time	Crew, repairs, insurance etc.	Fuel, port charges
Bareboat	—	Crew, repairs, insurance etc. Fuel, port charges

Similar basic forms of contract are used in the offshore hydrocarbon industry, in that an offshore vehicle may be employed to undertake a specific task (c/f Voyage Charter), or may be hired for a period of time.

The Owner's Options

Considering first the case of the tramp shipowner there are a range of options open to him regarding the employment of his existing vessels in terms of the lengths of charters. Charters for only a single voyage catering very much for marginal requirements show the greatest fluctuations in rates while the longer-period charter rates fluctuate more slowly and do not experience the same extreme highs and lows.

Thus a company owning a number of ships can spread its risks by placing a proportion of its fleet on long-term charters, although probably only at a very low profit, while chartering the remaining ships short-term. In this way a cash flow is ensured even when the market is very depressed, and ships without long-term charters are laid up while large profits can be generated when the market experiences a period of high rates. Considerable skill is required to charter a fleet to take maximum advantage of the periods of high rates without jeopardising the very existence of the fleet during periods of low rates.

Another area where skill is required is in the selection of possible charters that may be offered in order that the vessel is in a position at the end of one charter from which another loaded voyage can be arranged. If a charter has to be followed by a long ballast voyage profits will quickly disappear.

It will be appreciated from the above that when short-term rates make a sudden change it initially effects only a very few ships which happen to be coming off charter. Gradually, however, the effects of the change, if maintained, find their way right across the whole spectrum of shipping.

Sometimes shipowners can arrange Contracts of Affreightment with cargo owners for the carriage of a number of cargoes, not necessarily by nominated vessels. Indeed there was a time in the late 1960's when the modern bulk carrier had been evolved when many owners considered the future for this type of vessel lay in such contracts. Consortia were formed to enable large contracts to be obtained, but in the event it was found that this type of contract could usually be obtained only at low rates and the Consortia have largely faded away.

Specialised ships are very often chartered for long periods, five or ten year time Charters being frequently arranged, whereas bulk carriers are rarely chartered for more than three years. An exception is specialised vessels for the offshore industry where charters tend to be short, anything over three years being exceptional. The majority are for a year or less.

Owners who concentrate on specialised ships will be constantly on the lookout for new opportunities presented by new manufacturers or chemical processes requiring the bulk supply of materials, perhaps of a dangerous nature, and/or requiring specialised handling, and the construction of new industrial complexes. The changing patterns of trade as developing nations build up their own industries can also give rise to opportunities for specialised shipowners.

The owners of vessels employed in the offshore hydrocarbon industry keep in close direct touch with the operators in the industry and endeavour to anticipate their requirements both for existing types of craft, and also for new types that may be required by the ever-developing technology of the industry.

Sale and Purchase

As well as being concerned with the employment of his existing craft an owner will always be watching for opportunities, both making advantageous purchases of new units, and advantageous sales of existing units. Indeed as has often been said, owners can make larger profits from timely purchase and sale of units than can ever be made from operating them.

The prices of both new and secondhand ships are affected by demand particularly the latter where it is almost purely supply and demand that determines the price; diagram 13 shows a typical variation in the price of secondhand ships against a background of freight rates.

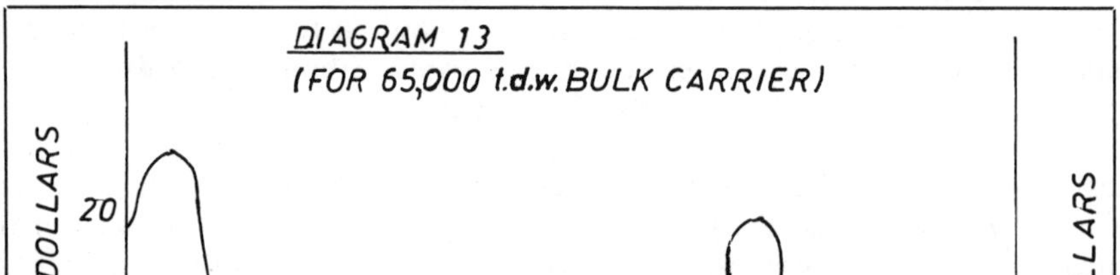

Diagram 14 shows new ship prices corrected to remove the effects of inflation, that is to say it shows the real cost of a particular size of ship over the period. The real cost will be increased by higher standards of accommodation for the crew; by more stringent regulations by such things as safety equipment; by more sophisticated navigational equipment, etc., and generally higher speeds requiring more powerful machinery. The real cost will be reduced by more economical methods of ship construction; improved structural design; improved design and constructural methods for the machinery; sometimes by changed regulations (for example, the Load Line Act 1966 allowed deeper loading), by smaller crews etc. It will be seen that in real terms there is still a very marked link between price and demand for new ships which is largely masked by inflation, but which may become more apparent to owners in the future as inflation accounting is introduced. The *Fairplay* 11/13,000 ton cargo vessel is the example used.

Other Factors

Shipping is essentially an international business, and a shipowner has to face competition from owners in many other countries. The same applies to the owners of craft in the offshore hydrocarbon industry, except that there are fewer countries involved, and the craft tend to be of a more specialised and sophisticated nature. International agreements on safety standards have been in force for many years ensuring that competitors all maintain at least the same minimum standards as will be seen from the next chapter. Co-operation between Trades Unions in different countries is leading to certain minimum standards of pay, but there are nevertheless very large differences between the costs of providing a European crew compared to say a crew from the Far East, and this is one of the problems facing shipowners in the traditional maritime nations.

Another problem which such owners have to face is that of flag discrimination which upsets the usual commercial forces by diverting cargoes to ships of a particular nation which may be inefficiently operated. Many nations now exercise flag discrimination in one form or another by various means, in both the shipping and offshore hydrocarbon industries, forcing cargoes into ships of their own flag. This presents obvious difficulties for owners operating liner services to that country, and to tramp shipowners it leads to more ballast voyages, thus in both cases leading to uneconomic operations.

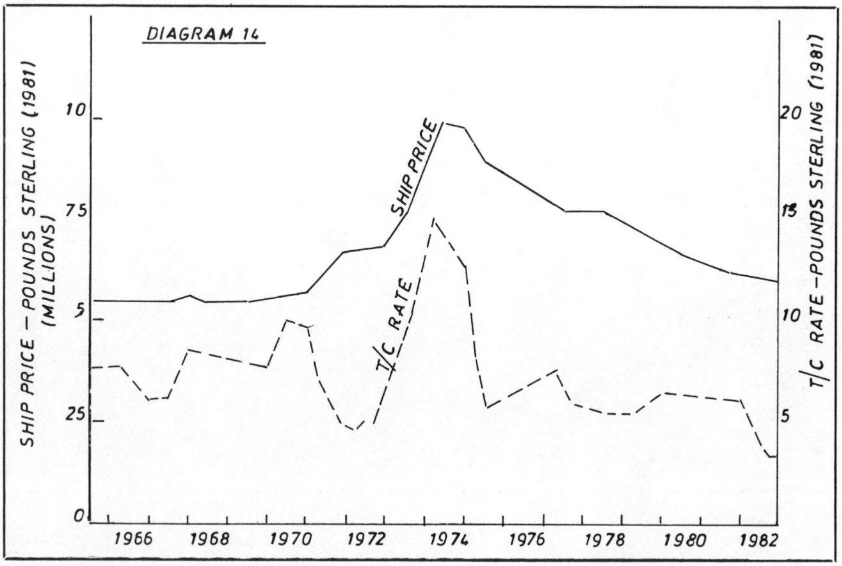

Shipbrokers

The reader may well wonder how the shipowners make contact with their charterers, bearing in mind the enormous size of the international shipping industry. The answer is through brokers operating very largely on the Baltic Exchange in London. Some shipping companies have their own chartering staff working at the Exchange, while others may employ brokers, known as Owners Brokers. Merchants and other interests with cargo to be moved employ brokers known as Chartering Agents. There are many broking firms who number both owners and charterers among their clients.

As well as arranging chartering operations a typical shipbroking firm will also undertake to arrange the sale and purchase of ships for its clients, including both new and secondhand vessels. Other shipbrokers located at ports throughout the world undertake ships agency business, that is to say, they provide services to ships visiting the port that cannot conveniently be provided by the owner from a distant office, for example assisting with Customs clearance; dealing with local suppliers, etc. Other shipbrokers may be employed by the charterers of the ship to look after their interests when, for example, a ship on charter enters a port to load cargo.

The offshore market, although growing rapidly, involves far fewer owners and charterers, and a large proportion of the contracts for the employment of vessels and rigs are arranged direct between owner and charterer or hirer. Some shipbroking companies have diversified into offshore activities, while new broking firms have been set up by experienced offshore personnel. As the size of world offshore market increases so will the opportunities for brokers.